AF654891

PRINCIPLES OF AIRFLOW FOR NATURAL VENTILATION

A GUIDE FOR STUDENTS OF ARCHITECTURE AND ENVIRONMENTAL DESIGN

AKUBUE JIDEOFOR ANSELM

BlueRose ONE.com
Stories Matter
NewDelhi • London

BLUEROSE PUBLISHERS
India | U.K.

Copyright © Akubue Jideofor Anselm 2025

All rights reserved by author. No part of this publication may be reproduced, stored in a retrieval system or transmitted in any form or by any means, electronic, mechanical, photocopying, recording or otherwise, without the prior permission of the author. Although every precaution has been taken to verify the accuracy of the information contained herein, the publisher assumes no responsibility for any errors or omissions. No liability is assumed for damages that may result from the use of information contained within.

BlueRose Publishers takes no responsibility for any damages, losses, or liabilities that may arise from the use or misuse of the information, products, or services provided in this publication.

For permissions requests or inquiries regarding this publication, please contact:

BLUEROSE PUBLISHERS
www.BlueRoseONE.com
info@bluerosepublishers.com
+91 8882 898 898
+4407342408967

ISBN: 978-93-6783-066-6

Cover design: Yash Singhal
Typesetting: Namrata Saini

First Edition: March 2025

Acknowledgement

My sincere appreciation goes to my family whose support over the years provided me the break to do this book. I also extent my special gratitude to Dr. Adekoyejo B. Jolaoso (Rector Moshood Abiola Polytechnic, Abeokuta), whose reviews and constructive criticism assisted in the revisions that aided in getting to the final copy of this book.

Akubue Jideofor Anselm

Baze University Abuja, 2024.

Foreword

This book is a concise and accessible introduction to the fundamental principles of airflow in relation to natural ventilation in the designs of buildings. The book is a collection of the author's teaching, research and practical experiences that is capable of inspiring, enlightening and inculcating the better understanding of the subject.

The book is a timely response to rising quest for reference materials from students in the schools of environmental studies and admirers of architecture and sustainable built environment. The book presents illustrative insights into the subject. It also demonstrates the ability and capability of stimulating students, upstarts and the general readership's better understanding of the principles, importance and practicability of natural ventilation and passive cooling for a sustainable energy saving in the evolution of building designs that; otherwise misconstrued and considered misplaced in the understanding of the inherent issues involved, especially in residential buildings.

This nine chaptered book commences with a brief introduction of the subject, followed by the discussion of the underlying principles, types, comparisons, challenges and disadvantages of ventilation; application in building designs and impacts of airflows regimes on buildings. Furthermore, it presented the design strategies and touches on the practical demonstrations with explanatory illustrations and case studies.

The book thoughtfully concluded with a justification and importance of the book towards generating indigenized/localized evaluation systems with the appropriate description of airflows, thermal conditions and natural ventilation outlooks within tropical regions in an integrated manner. This book is no doubt, an interesting expository towards environmentally friendly and energy efficient buildings.

The initiative and efforts deployed in writing this book is highly commended, which is bound to be of immense benefit for the advancement of knowledge in the subject area. I therefore, recommend the book as a useful and relevant study material amongst students, researchers, practitioners, policy makers across all Built Environment disciplines and general readership.

Adekoyejo B. Jolaoso (PhD)

Abeokuta Ogun State, Nigeria. 2025

Prologue

Different definitions are used for description of the term ventilation. While some definitions are driven by natural processes, others are driven by technical applications.

A typical definition from Cambridge English Dictionary simply implies that ventilation is the movement of fresh air around a closed space[1]. Other definitions include the intentional introduction of outdoor air into a space, the process by which 'clean' air (normally outdoor air) is purposely provided to a space and stale air is removed by either natural or mechanical means[2]. Also from the medical perspective, ventilation is defined as the movement of outdoor air into a building or a room, and its distribution within the building or room. According to this definition, the general purpose of ventilation in buildings is to provide healthy air for breathing by both diluting the pollutants originating in the building and removing the pollutants from it[3].

In architecture and related disciplines, ventilation is principally viewed as the means of air circulation inside living spaces. Within the indoor environment, ventilation is mainly used to control air quality by diluting and displacing indoor pollutants as well as to control indoor temperature, humidity, and air motion for the benefit of thermal comfort and satisfaction of occupants.

The benefits of indoor ventilation include the expulsion of build-up of pollutants, bacteria, moisture and unpleasant odors as well as airflow regulation. While the benefits of indoor ventilation abounds, the benefits of natural ventilation

---

[1] *Cambridge Advanced Learner's Dictionary*

[2] *AIVC (1996). "GV: A Guide to Energy Efficient Ventilation"*

[3] *D.W. Etheridge, M. Sandberg, A simple parametric study of ventilation, Building and Environment, Volume 19, Issue 3, 1984*

on the other hand is mostly consigned to the context of energy efficiency.

For indoor ventilation to occur there must be air movement. Ventilation is driven by either natural wind-flow or by mechanical means. Natural ventilation or otherwise referred to as Passive ventilation is the process of supplying air to and removing air from an indoor space without using mechanical systems. Over the years, the significance of natural/passive ventilation has become prominent especially in tropical regions like Nigeria, owing to its benefits in reducing building energy consumption and environmental sustainability. The natural drivers of ventilation are discussed in this book as a way of broadening students' understanding of the principles of wind-flow for the purposes of effective application of natural ventilation in building designs.

The idea of putting together this book was driven by the interest to expand students' understanding of the processes of natural ventilation and passive cooling. Students are often tasked with presenting building designs that incorporate sustainable ideas and energy saving prospects. However, the products seen at the end of their projects signifies low understanding of the fundamental principles of wind flow within the built environment in tropical regions such as Nigeria. In the light of these, this book initiates the adoption of passive concepts for the purposes of implementation and evaluation of natural ventilation principles in residential buildings. It is hoped that bringing this book to light would further broaden the minds of student architects in their quests for developing more environmentally friendly and energy efficient buildings.

Contents

CHAPTER ONE

Principles of Ventilation

Ventilation has been identified as the process of moving outdoor air into a building or space and its distribution the within the building or space. Circulation thus refers to the flow of the air, which is a significant feature of both open and closed systems. In fundamental terms, there exist two basic processes of a ventilation system which work in different ways. The first is in an open system, while the second is in a closed system. Ventilation works in conjunction with Air exchange, which is the physical exchange of air-mass in one space, with a new air-mass introduced external as replacement to that same space.

The flow rate of ventilation either referred to as:

a. Absolute ventilation flow rate in *1/s* or m^3/s, or
b. Air-change rate relative to the volume of the space.

Measured in cubic meters, the ventilation rate is referred to as the absolute amount of inflow air per unit time, i.e. litre per second or 1/s, cubic meter per hour or m^3/hr. The air-change rate on the other hand is referred to as the relative amount of inflow air per unit time.

The relationship between ventilation rate in *1/s* and air-change rate is expressed by the equation[4]:

Air-change rate = *[ventilation rate (l/s) × 3600 (s/hr)] × 0.001 (m^3/s)]/[room volume (m^3)] Eq.1*

or,

---

[4] *World Health Organization. (2009). Natural ventilation for infection control in health care settings. World Health Organization.*

Ventilation rate (l/s) = air-change rate × room volume (m^3) × 1000 (l/m^3)/3600 (s/hr)   Eq.2

In the built environment, Ventilation performs two basic functions, which are air exchange and air distribution.

Air exchange

Air exchange system is classified as an open ventilation system. It is defined as the cycle of fresh air in and stale air out of a space. Air exchange ensures that indoor environments are kept free from pollutants as well as maintains cooler spaces by removing stale indoor air while recovering the cooler and fresher air to sustain indoor comfort quality. For air exchange to occur there must be a fixed *inlet* source and an *outlet* (exhaust). Most natural or passive ventilation processes are considered as open systems.

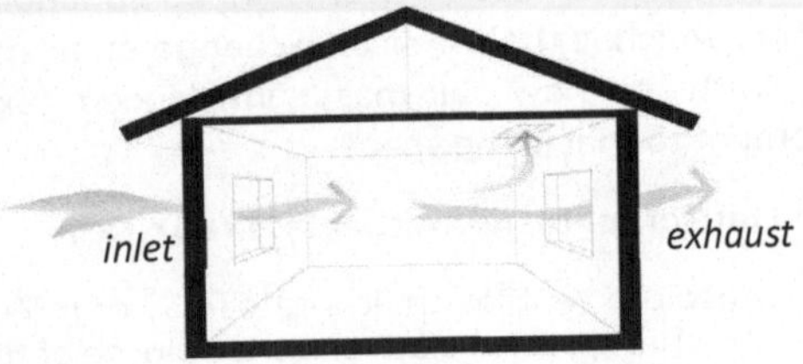

Fig. 1. Air exchange (open process)

Air distribution

Air distribution of circulation is defined as the process of delivering fresh air to all spaces in a building and the mixing of fresh air with stale air prior to its removal from a building. This process is mostly achieved by mechanical or hybrid ventilation modes. The processes of air circulation can either be open system based or as a closed ventilation system. The combination of fresh air *inlet* (source), distribution and exhaust (*outlet*) system are utilized in air distribution process (Fig. 2). While use of Inlets provides the primary means of controlling air distribution within a ventilation space, mechanical controls systems like fans, inlets, ridge vents, and sidewall openings are utilized to achieve the desired circulation and ventilation performance within the space.

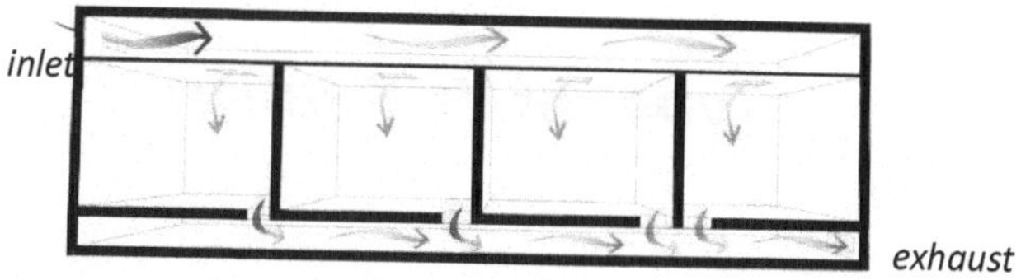

Fig. 2. Air distribution (closed process)

CHAPTER TWO

Types of Ventilation

The two primary types of Ventilation used in Building designs are mechanical and natural. While Mechanical ventilation utilizes mechanized fans to activate airflow, Natural ventilation on the other hand takes advantage of the naturally occurring wind forces to move air into and out of buildings. In either of these two types of ventilation, pressure difference causes air to flow, thereby providing the driving force for ventilation[5].

Mechanical ventilation:

As earlier stated, it takes mechanized fans to drive mechanical ventilation process. These fans can be installed either directly in windows (on walls), or installed within air ducts for air supply or exhausting air from room spaces.

Mechanical ventilation systems depend majorly on inclement climate. In warm and humid climates where condensation need to be reduced, infiltration may need to be prevented, in which case a positive pressure mechanical ventilation system is applied. The reverse is the case in cold climates, where exfiltration needs to be prevented to reduce condensation, in which case negative pressure ventilation is often adopted. A typical case is where locally generated pollutants occur in indoor spaces like bathrooms, toilet or kitchen, in which case also, negative pressure system is utilized.

Positive pressure systems are also operated as open systems, whereby indoor spaces experience positive pressure, leading

---

[5] *Mahmood, Farhat & Al-Ansari, Tareq. (2023). Renewable Energy Driven Sustainable Greenhouse: An Overview.*

the air in these spaces to be drawn out through envelope openings. Conversely, closed systems refer to negative pressure systems, in this case, the indoor spaces experience negative pressure, and the indoor air is compensated by mechanically drawing in air from outside[6].

Mechanical ventilation systems are applied in buildings where the necessity for improved ventilation is desired. Below are some of the advantages of this system.

- Mechanical ventilation systems are considered reliable in delivering desired flow, irrespective of the impacts of wind, humidity and ambient temperature. This system can be integrated easily into indoor air-conditioning with controls for the indoor air temperature and humidity.
- Air Filtration systems can be enabled in mechanical ventilation to keep off harmful microorganisms, particulates, gases, odors and vapors.
- Airflow paths in mechanical ventilation systems are adjustable, such that the air may be configured to flow from areas where there is a source of infection towards the areas free of occupants.
- Mechanical ventilation can work everywhere when electricity is available.

Challenges and disadvantages of operating mechanical ventilation systems are likewise listed as:

- Installation and maintenance costs for mechanical ventilation system are often quite high. In the case installation and maintenance of mechanical system are affected by lack of funds, its performance will be adversely compromised.

Natural ventilation:

Owing to indoor and outdoor air density differences, natural forces like winds and thermal buoyancy drive outdoor air through openings in building envelopes causing Passive air

[6] *Sehulster, Lynne & Chinn, Raymond. (2003). Guidelines for Environmental Infection Control in Health Care Facilities. Recommendations and reports / Centers for Disease Control.*

flow which is the primary driver of natural ventilation. These building envelope openings include windows, doors, chimneys, and other forms of passive ventilators. The natural ventilation processes depends hugely on climate factors and building design factors (such as building orientation).

Passive or natural ventilation systems when compared with mechanical ventilation systems have numerous advantages especially in tropical less economically developed regions. Some of the advantages are mentioned below.

- Natural ventilation when properly done can generate high ventilation flow rate cheaper economically than mechanical systems, due to the use of natural wind driven forces and openings.
- It is considered to be more energy efficient in tropical environments where heating is not required.
- Technologically, natural ventilation may be classified into simple natural ventilation systems and high-tech natural ventilation systems. In the case of the high-tech system which is computer-controlled, the system may be assisted by mechanical ventilation systems thereby referred to as hybrid or mixed-mode systems.
- Though the High-tech natural ventilation system may have the same limitations as mechanical ventilation systems, it however harbors both benefits of mechanical and natural ventilation systems, thereby improving its efficiency value.
- Generally, the advantage of natural ventilation lies in its ability to provide a very high air-change rate at very low cost controlled by a very simple system which can achieve high air-change rates by natural wind driven forces.

The disadvantages of natural ventilation system include:

- Its dependence on outside climatic conditions relative to the indoor environment. This is owing to the two driving forces that generate the airflow rate (i.e. wind and temperature difference). Owing to this, Natural ventilation may be difficult to control, with airflow being uncomfortably high in some locations and stagnant in others.

- The other challenge lies in the difficulty in controlling the airflow direction due to the absence of a well-sustained negative pressure.
- Natural ventilation has been identified to exclude the use of particulate filters. Also, elements like climate, security and cultural criteria often impact efficiency of natural ventilation caused by the closure of windows and vents, which eventually result to lowering of ventilation rates.
- Natural ventilation has also been identified to function only when natural forces are available.
- Natural ventilation systems are often affected by poor design, poor maintenance or inappropriate management.
- Application of natural ventilation requires that fenestration system be left open leading to other discomforts such as noise, air pollution, insect vectors and security challenges.

From the descriptions above, the difficulties identified in the operation of natural/passive ventilation can be overcome by adoption of hybrid ventilation options. Table 1 presents the summary of advantages and disadvantages of the different types of ventilation systems discussed in this book.

Table 2.1Summary of advantages and disadvantages of different types of ventilation systems.

Value	Mechanical ventilation	Natural ventilation	Hybrid ventilation
Advantages	Suitable for all climates and weather with potential for air-conditioning as climate dictates	Most suitable for warm and tropical climates with 50% of natural ventilation attainable	Suitable for most climates and weather
	Better controlled and creates comfortable environment	Lower capital, operational and maintenance costs for simple natural ventilation	Energy-saving
	Enables lesser range of control of environment by occupants	Capable of achieving high ventilation rate and large range of control of by occupants.	Flexible
Disadvantages	Expensive to install and maintain	Easily affected by outdoor climate and/or occupant's habits and culture	Moderately expensive
	Reported failure rate in delivering the required outdoor ventilation rate	Difficult to predict, analyze and design	Difficult to design
	Potential for noise from equipment	- Reduces comfort level of occupants due to change in climate such as hot, humid or cold conditions. Inability to establish negative pressure in isolation areas. - High potential for insecurity, noise and insect intrusion. - High-tech natural ventilation shares almost same limitations and disadvantages as mechanical ventilation	

CHAPTER THREE

Natural Ventilation Principles

Findings from research reveals that the use of outdoor air flow for natural ventilation, combined with natural cooling and day-lighting techniques, have been essential elements of architecture since historic times, stemming from the first part of the 20th century[7]. Classical architectural design made with H, L, T or U-shaped floor plans were adopted within open courtyards, with the application of maximum windows sizes for the purposes of exploiting natural ventilation. Likewise in recent times, natural ventilation is greatly replacing mechanical ventilation systems in the high and middle-income regions all across the globe.

References for building standards relating to ventilation rates, in developed countries like the United States refer to ASHRAE Standard 55-2010 (Thermal Environmental Conditions for Human Occupancy[8]). The scope ASHRAE Standard 55-2010 is significant for its objective which is to specify the combinations of indoor thermal factors for environmental and personal values that will produce acceptable thermal environmental conditions for majority of occupants within a space. The revised standard of 2004 however indicated that there are differences between naturally and mechanically conditioned spaces with regards to factors like occupant thermal response, change in clothing types, availability of ventilation control and shifts in

---

[7] *ASHRAE. Handbook of HVAC applications. Atlanta, USA: American Society of Heating Refrigerating and Air-Conditioning Engineers Inc.; 2007. Health care facilities.*

[8] *ANSI/ASHRAE Standard 55-2010: Thermal Environmental Conditions for Human Occupancy". Atlanta, GA: American Society of Heating, Refrigerating and Air-Conditioning Engineers. 2010*

expectations of occupants[9]. With regards to the Natural ventilation, an addition to the standard (5.3: Optional Method For Determining Acceptable Thermal Conditions in Naturally Ventilated Spaces), added adaptive thermal comfort approaches for naturally ventilated and conditioned buildings by adding acceptable operational temperature ranges for naturally ventilated and conditioned spaces. These applications by ASHRAE targeted towards the design of natural ventilation systems became significant in furthering the improvement of occupant-friendly spaces as well as contributing in evolving sustainable and energy efficient building designs.

The essential principle that enables the realization of natural ventilation is discussed in this section. The major principle under which natural ventilation occur in indoor spaces includes the following:

1. Passive Ventilation
2. Cross Ventilation
3. Stack Ventilation

1. Passive Ventilation

A general term used in describing ventilation, whereby no active assistance is utilized to achieve the process of air supply into spaces and its removal. Often used to represent natural ventilation, there are two types of passive ventilation occurring in buildings:

- Wind driven ventilation
- Buoyancy-driven ventilation.

i. Wind driven ventilation:

This form of ventilation is classified as cross ventilation or single-sided ventilation. This is owing to the fact that it is dependent on wind behavior and its interactions with

[9] *de Dear, Richard J.; Gail S. Brager (2002). "Thermal Comfort in Naturally Ventilated Buildings: Revisions to ASHRAE Standard 55". Energy and Buildings. 34 (6): 549–561.*

building envelope with openings often referred to as inlets or wind-catchers.

ii. Buoyancy-driven ventilation:

This type of ventilation is that which is formed due to differences in density of air, which in large part arises from differences in temperature. In the occurrence of temperature difference between two adjoining volumes of air, the warmer air naturally assumes a lower density and thus becomes more buoyant thereby rising above the cold air creating an upward stream. This system often creates the premise for Passive stack ventilators to be used in improving ventilation.

Whereas wind driven ventilation are generated from the different pressures created by wind around a building and goes into the building spaces through openings (windows and doors) formed on the perimeter of the building. Buoyancy-driven ventilation on the other hand, occurs as a result of the directional buoyancy forces resulting from temperature differences observed between the interior and exterior environments[10]. An illustration of the wind driven and buoyancy driven ventilation process is shown in Fig. 3 and 4.

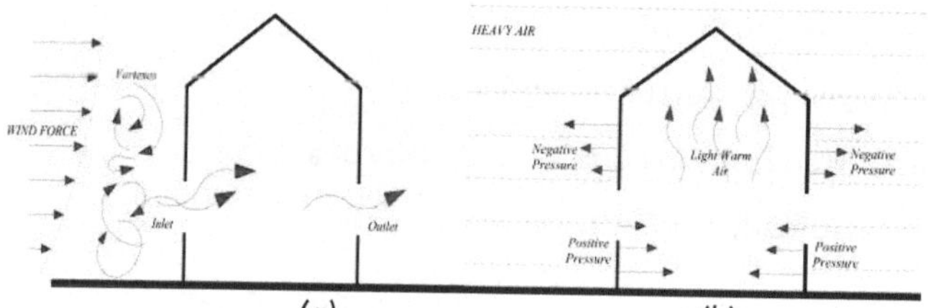

(a) (b)

Fig. 3. (a) Wind driven ventilation, (b) Buoyancy driven ventilation

[10] *Linden, P. F. (1999). "The Fluid Mechanics of Natural Ventilation". Annual Review of Fluid Mechanics. 31: 201–238*

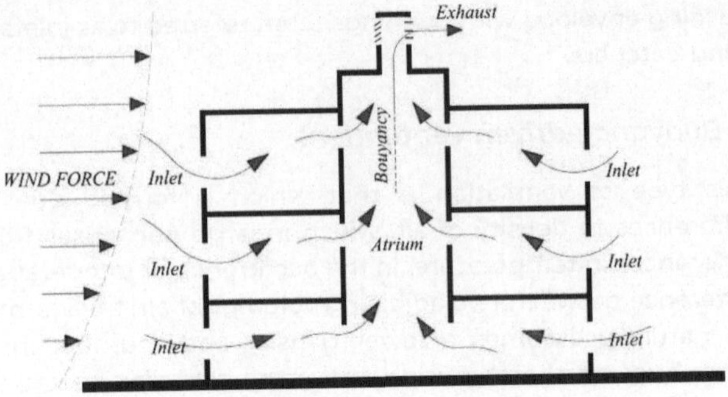

Fig. 4. Combined Wind and Buoyancy driven ventilation

The differences seen in the static pressure arising from global and microclimate thermal phenomena collectively contribute in creating the air flow referred to as "Wind". The pressure that is created when wind comes into contact with objects like hills or buildings are referred to as Dynamic pressures. These pressures are defined by the equation 3 below[11].

$$q=\frac{1}{2}\rho v^2, \qquad \text{Eq.3}$$

where (using SI units):

q = dynamic pressure in Pascals

ρ = fluid density in kg/m^3 (e.g. density of air)

v = fluid velocity in m/s.

The direct impact of wind on buildings affects the rates of ventilation and infiltration through it. This in turn impacts on the related heat losses or heat gains in the interior spaces. The actions of Passive ventilation often culminate in passive cooling, which is resultant process for indoor thermal comfort. When wind speed increases with height it is lowered towards the ground due to frictional drag forces, which vary significantly thereby creating complex air flows and turbulence when it interacts with elements found in the

[11] *Clancy, L.J. (1975). Aerodynamics. John Wiley & Sons*

environment (be it natural elements like trees and hills, or physical street/urban geometries like street light and landscaping furniture). In the past, traditional (vernacular) buildings were designed to rely majorly on passive ventilation for maintaining thermal comfort conditions in the enclosed spaces; however these concepts have become modern ideas for sustainable and green designs especially in poorer regions of the world.

2. Cross Ventilation

The interior climate conditions in houses within hot and humid climates like the tropics often times turn extremely hot and unbearable. In these circumstances, the process of cross ventilation can become one of the most effective ways of naturally cooling houses in the tropical regions. With the natural occurrence of a wind-driven strategy, cross ventilation utilizes wind or airflow to move air through room spaces, thereby reducing the temperature significantly.

The major propagator of natural ventilation is the wind-flow. The flow of wind across human living environments differs significantly from region to region dependent on climatic dispositions.

Cross ventilation is best applied in regions or locations that have ideal access to winds or breezes. In the presence of breezes are, the process of a cross flow ventilation becomes the easiest means to naturally remove stale air from internal room spaces.

The effect of Cross ventilation cannot be compared to that of a Single-sided one. In the case of single-sided ventilation, where only one side of a room has an opening (door or a window), the rate of air-exchange is reduced significantly, as the air movement has to return through the same inlet source for active exchange to occur. Below are the basic principles necessary for effective Cross ventilation in houses:

i. Minimum of two opening:

Cross ventilation works best when there is at least one opening facing prevailing winds and another opening on the opposite side of the room space as shown in fig. 5.

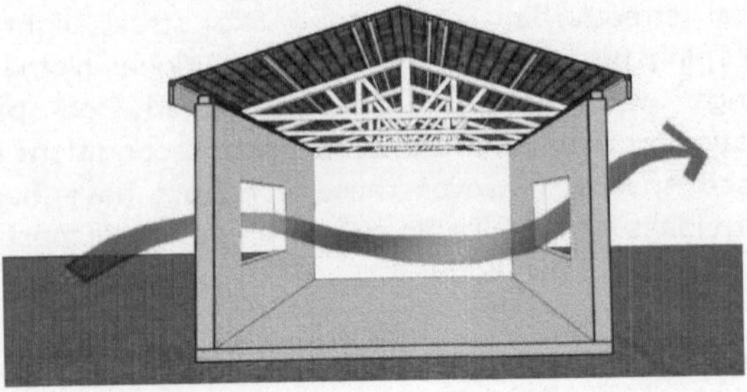

Fig. 5. Illustration of Cross ventilation with windows at opposite sides of wall

ii. Shorter distances between openings:

As the effectiveness of cross ventilation depends on the travels time of winds from one side of the room to the other, its performance would be better with reduced travel distance. In this case, as a room deepens in its depth (as demonstrated in fig. 6), the wind force slows exponentially and thus become less effective. Ideally, for the performance of the cross ventilation to be more significant, the distance between the openings should be as short as practically possible. Research suggests a range of distance not more than 12m for effective cross ventilation.[12]

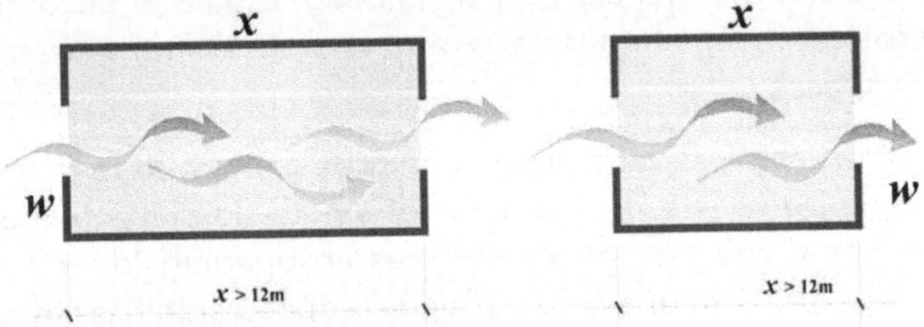

Fig. 6. Comparing effects of Cross ventilation in shorter and deeper room depths.

---

[12] *Hong Kong Government Buildings Department 2014.*

3. Stack Ventilation

Stack ventilation operates on the natural principles of temperature differentials to move air. Naturally, hot air rises owing to its lower pressure (referred to as buoyancy ventilation). Hence the stack effect order is described as the rise of hot air due to buoyancy, resulting in lower pressure, which consequently sucks in fresh air from outside.

As buildings often have open areas at lower levels (especially taller buildings), the stack effect occurs leading from the air infiltration from these lower level openings going upwards. This often is seen during the hotter seasons, when the warmer indoor air rises up through the building and tends to escape at the upper levels either through windows, ventilation openings, or non-intentional openings in ceilings or roofs. This rising warm air then reduces the pressure in the base of the building, thus drawing colder air inwards through these open doors, windows, or leakages from other openings as mentioned above. However, during the colder seasons, the stack effect itself reversed in buildings, though this is often weaker owing to lower temperature differences.

Stack ventilation is often referred to under the Bernoulli's principle of passive air movement. These two classes of passive ventilation largely depend on air pressure differences that rely on heights to pull air through the building. The impact of lower pressures that is naturally higher in the building aid in the suction of air upward.

Bernoulli's Theorem

Bernoulli's principle is a key concept in the field of fluid dynamics that relates pressure, density, height and speed in defining air movement. Bernoulli's principle states thus, *"An increase in the speed of a parcel of fluid occurs simultaneously with a decrease in either the pressure or the height above a datum"*[13]. This is as seen in the book *Hydrodynamica* published by Daniel Bernoulli in 1738, which presented theories on fluid dynamics. According to Bernoulli's principles on energy conservation, *"as the speed*

[13] *Clancy, L.J. (1975). Aerodynamics. John Wiley & Sons*

of a moving fluid increases (liquid or gas), the pressure within the fluid decreases"[14], this principle thus identifies with the airflow effects experienced in stack ventilation.

Though this principle is often applied in building design, in architectural terms, outdoor air which is above the ground is mostly less obstructed, hence it moves quicker than the air that is lower, thus maintains lower pressure. The resultant lower pressure is what then sucks fresh air through the building. This effect though effective in architectural terms, can be largely influenced by a building's surroundings due to the number of obstruction within the environment. Air movement (wind) is controlled by three major forces near the earth's surface[15] which are:

i. force due to pressure gradient,
ii. deflective force in relation between pressure and wind field (the Coriolis force)
iii. force due to the law of angular momentum.

When winds (air in motion) strike a building at right angles, it would exert pressure on the face of the building. Consequently, the wind is deflected over and round the sides of the building resulting in fields of relatively reduced pressure. This resultant (induced) pressure differential that is created across the building is what causes indoor airflows (passive ventilation).

Dependent on its velocity, wind (air in motion) has a pressure given by the Bernoulli's equation (*Clancy, 1975*) as follow:

Wind pressure:

$$P_W = \tfrac{1}{2} p\, v^2 \qquad \textit{Eq.4}$$

Where p = air density,
v = velocity in ms-1

---

[14] *Danielis Bernoulli (1738). Hydrodynamica, sive de viribus et motibus fluidorum commentarii : opus academicum ab auctore, dum Petropoli ageret, congestum*

[15] *Markus, T.A. and Morris, E.N. (1980) Buildings, Climate and Energy. Pitman Publishing Limited, London.*

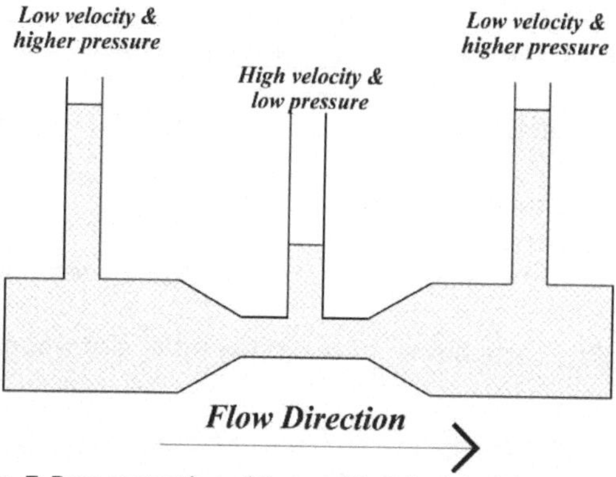

Fig. 7. Demonstration of Bernoulli's Principle (Clancy, 1975)

Chimney effect

Stack effect is also referred to as Chimney effect. This is due to the movement of air through a building resulting from buoyancy, via building elements like chimneys, flue-gas stacks and other purposefully designed openings or containers. The similarities between the stack effect experienced in taller buildings and that of Chimneys or industrial flue gas stacks are almost identical, except that in the case of flue-gas stacks and Chimneys, it involves hot flue gases having large temperature differences with the ambient outside air. Furthermore, higher temperature differences between the outdoor air and the flue gases creates strong stack effect in building elements like Chimneys for buildings using a fireplace for heating, thus presenting the significant reference termed *"Chimney effect"* in stack ventilation.

The potentials of stack-effect ventilation can be estimated for design processes through the use of the equation[16] below:

V = *60 K A [g h (Ti –To)/Ti]0.5*   *Eq.5*

Where;

[16] *Abd W. I., Ismail, L. & Kadir, A. (2016). Opening Design and Position Effect on Building Natural Stack Effect and Cross Ventilation. International Journal of Engineering and Technical Research.*

V is the estimated air flow rate (cfm);
K is a factor that accounts for orifice characteristics (assumed = 0.65);
A is the smallest value of inlet, stack "throat," or outlet areas (m^2);
g is the gravitational constant (9.8 m/sec^2);
h is the stack height (m);
Ti (inside) is the higher of two differential air temperatures (°R);
To (outside) is the lower of two differential air temperatures (°R).

In the process of design, three major factors affect the rate of stack Ventilation, which are[17]:

- Area of openings (i.e. inlet, outlet, and the stack size).
- Differences in the elevational height between the inlet and the outlet.
- Difference in the dry-bulb air temperatures between the stack inlet and the outlet.

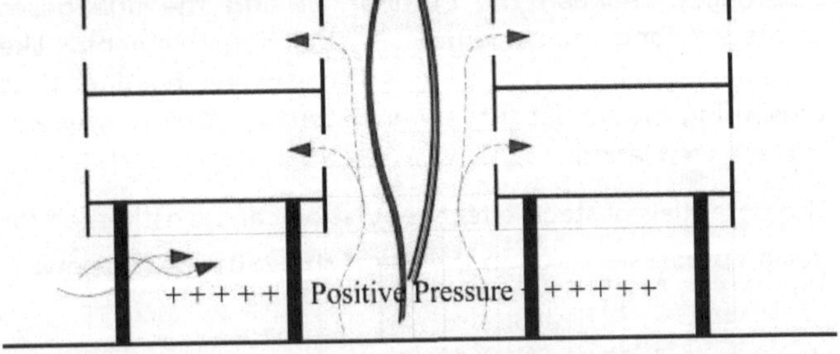

Fig. 8. Combination of Bernoulli's principle, Chimney/Stack effect in passive ventilation.

---

[17] *Abd W. I., Ismail, L. & Kadir, A. (2016). Opening Design and Position Effect on Building Natural Stack Effect and Cross Ventilation. International Journal of Engineering and Technical Research.*

In real application, the two principles of stack and Bernoulli have similarities and differences and they often play out simultaneously in passive ventilation of buildings as shown in Fig. 8. The major advantage of the Bernoulli's principle over the stack effect lies in the fact that it multiplies the efficiency of wind ventilation. On the other hand, the stack ventilation has a major advantage over Bernoulli's principle in the fact that it does not need wind flow to be initiated, rather it works well on still (windless days) days when it may be most needed in tropical areas. In architectural application, designing for one of the two effectively represents designing for both, as the strategies may be the same.

CHAPTER FOUR

Application of Natural Ventilation in Building Design

Ventilation originated from the Latin word "*ventilation*", or "*ventulus*" meaning a breeze. The earliest practice of natural ventilation was seen in the roman "hypocaust system" (Fig. 9), which was a system of central heating that produces and circulates hot air below the floor of a room, as well as the walls with series of pipes and tile vents through which the heated air is distributed[18]. Historically, the earliest reference to the application of this system was by Vitruvius, who attributed its invention to *Sergius Orata* in c. 80 BC.[19] Its application was also seen in the Temple of Ephesus in 350 BC. The invention and adoption of this system was generally seen to improve the hygiene and living conditions of citizens in those earlier days and thus became the pioneer of modern central heating system in buildings[20].

---

[18] *Tomlinson, Charles (1850). A rudimentary treatise on warming and ventilation: being a concise exposition of the general principles of the art of warming and ventilating domestic and public buildings, mines, lighthouses, ships, etc. J. Weale. pp. 53. hypocaust.*

[19] *Forbes (1966-01-01). Studies in Ancient Technology*

[20] *Mitchell, Patrick (2008-03-01). Central Heating, Installation, Maintenance and Repair. Writers Print Shop.*

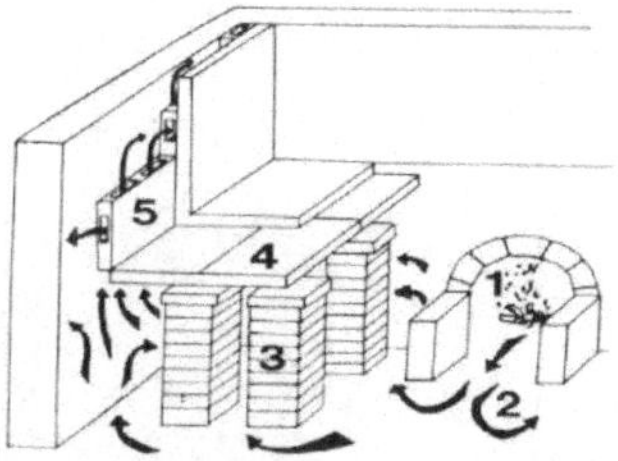

Fig. 9. Schematic representation of a hypocaust bath as represented (Source: Adam, 2005)[21]

(1) the furnace, (2) exhausted gas, (3) the pilae (pillars), (4) floor, (5) tubuli (pipes).

Later in region of the 1660's, the definition of ventilation was redefined as "the process of replacing foul air in an enclosed place with fresh, pure air", and this became the modern definition for Natural Ventilation.

Other significant traditional applications of natural ventilation are observed in building elements like Wind towers (Fig. 10). A wind towers (bagdirs) often referred to as windcatcher or wind scoop are distinctive and ancient feature of Persian/Islamic architecture used to create cross ventilation and passive cooling in buildings. Designs of wind towers can be found throughout the Middle Eastern region, including places like Pakistan and Afghanistan, though presently can be found incorporated into Western architecture[22].

---

[21] *Adam, J. P. (2005): Roman building: materials and techniques. Routledge.*

[22] *Yang, Tong & Clements-Croome, Derek. (2012). Natural Ventilation in Built Environment.*

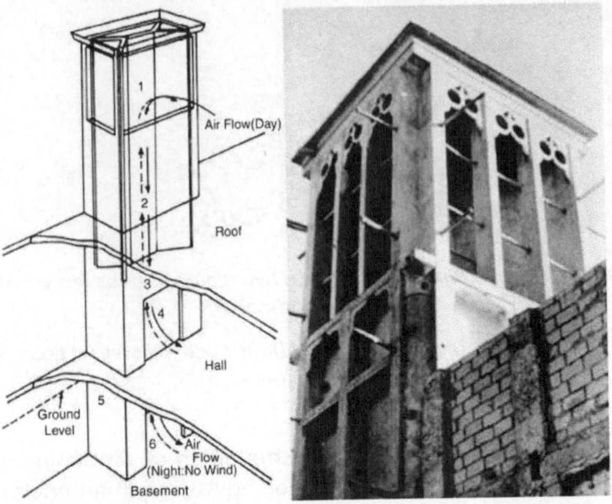

Fig. 10. Image of a Wind tower (Bagdir) as seen in Dubai, UAE[23].

Traditional Windcatchers function in two major ways:

i. By guiding airflow into it, using the pressure of wind force,
ii. By guiding airflow into it using the buoyancy forces from temperature gradients.

Another traditional element used for applying natural ventilation in buildings is the Courtyards. These are described as circumscribed areas (circular or square), that is open to the sky and often surrounded by building units. Courtyards are open areas surrounded by walls of buildings as seen in Fig. 11. In building design, they are special voided elements typically located in a building's center to serve different functions.

Courtyards are commonly seen in both Western and Eastern architecture and have been used in both ancient and contemporary buildings as typical building feature. Courtyards first appeared in the Neolithic Yarmukian site at Sha'ar HaGolan, in the central Jordan Valley in c. 6400–6000

---

[23] *Fathy H (1986) Natural energy and vernacular architecture, principles and examples with reference to hot arid climates. The University of Chicago Press, Chicago*

BC along the northern bank of the Yarmouk River[24]. These architectural elements historically were reserved for functions like cooking, working, sleeping, playing, gardening and for keeping animals.

Fig. 11. Conventional courtyard setting in West Africa[25]

Courtyard application in buildings can be traced as far as the BC 6400 within the Jordan valley where homes were constructed around open squares spaces surrounded small room units. This practice gained prominence within the Ancient Roman and Chinese housing styles where central courts and gardens were built to be surrounded by homes for luxurious living. However, thls practice gained prominence within the warmer climate regions within the 20th century for purposes of bringing in sunlight within the living quarters and introducing aeration within interior living spaces. The advantages of courtyard buildings cuts across its potentials for introducing cross ventilation in room spaces and improvement of thermal comfort through air movement during the very hot seasons. This potential has remained the most significant value for the application of courtyards in building designs. Likewise, with the advent of energy saving and sustainable building development, it is seen as the most effective passive design strategy for improving thermal comfort within the tropical regions where high heat and

---

[24] *Garfinkel Y. 1993. "The Yarmukian Culture in Israel". Paléorient.*

[25] *https://livinspaces.net/design-stories/articles/making-a-case-for-the-courtyard-in-nigerian-contemporary-architecture-an-article-by-laolu-rahman-ganiy*

humidity abounds. Numerous studies have been conducted in the past to understand the effects and impacts the courtyard spaces pose on the building system.

Courtyards as an architectural design style have been incorporated into buildings. Apart from serving the purpose of central gathering court for family and other activities, its major function lies in its adoption as air-wells for fresh air intake as well as serving the function of light-wells for providing light source to surrounding spaces through its open sky features[26].

Different configurations of courtyards are adopted in different locations to serve a variety of functions ranging from social gathering areas to passive lighting and ventilation purposes. In the tropical Nigerian environment, the primary function of courtyards is that of passive cooling through natural ventilation. Studies of existing courtyard configuration all across Nigeria identified a variety of compositions which could be categorized as vernacular, colonial, post-colonial and modern. This concept of architecture is seen in various building types ranging from residential buildings to administrative, institutional, health, educational, and multi-functional buildings.

Courtyards are designed to function as wind catchers, as they trap flowing wind into the enclosed open space for air recirculation. This recirculated air is then used for passively ventilating the room spaces around the courtyard as seen in Fig. 12.

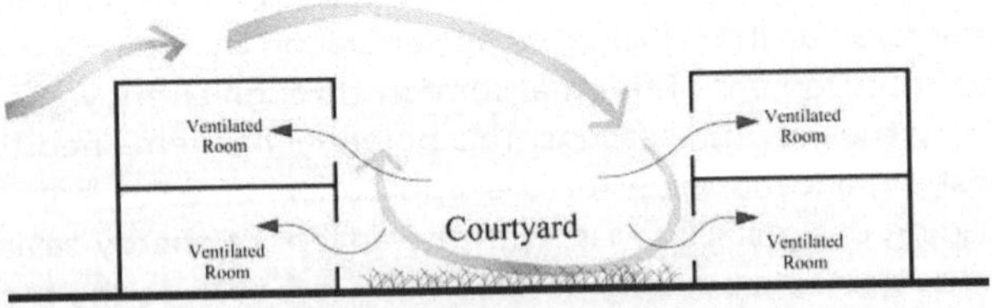

Fig. 12. Sketch of a typical courtyard showing the process of natural ventilation with air recirculation

[26] *Saxon. R. (1994).The Atrium Comes of Age. Longman United Kingdom.*

Generally, traditional building strategies that promote natural ventilation and cooling have evolved over the years as they are adapted over time by people from different climates and cultures to meet specific human needs. Despite these evolution and adaptations of its design an application, the modern application of these strategies still maintains the basic principles which position them as sustainable options for building design and services, based on their economic and energy saving qualities.

CHAPTER FIVE

Principles of Airflow for Natural Ventilation in the Built Environment

Natural ventilation for use in passive designs is hugely dependent on micro climate conditions of urban surroundings. Study indicates that this condition is affected by ambient wind flows, solar radiation and ambient air temperatures. In recent years, several studies on air circulations in streets have been performed by physical and numerical simulations. These studies focused on such parameters as street geometry (including height/width ratios), street patterns and orientation. Many researchers proposed that at the micro level, the urban space geometry is mostly responsible for the microclimate difference. Those responsible for the design of urban spaces are mostly unfamiliar with the thermal comfort problems originating from the distinctive relationship between urban forms and local climatological conditions[27]. Consequent to this, many urban settlements do not respond to related climatic conditions. The common practice for urban and building designers involves the use of universal meteorological data. However, these data are gotten from stations that are located in unobstructed areas of the city like the airports or isolated fields. Previous studies have shown that the use of inappropriate wind and temperature data in the design of urban buildings may result in inaccuracies regarding the air

---

[27] *Gomez, F.; Perez, A., Valcuende, M., Matzarakis, A. (2013). Research on ecological design to enhance comfort in open spaces of a city (Valencia, Spain). Utility of the physiological equivalent temperature (PET).*

flow in urban buildings[28]. Investigation into urban air flow identified the structure of urban wind field as characterized by the development of a two layers vertical structure. As air flows from the rural to the urban environment it must adjust to the boundary conditions defined by the cities, and these results to the development of the urban canopy sub-layer (obstructed sub-layer) which is extended from the ground surface up to the buildings height, while the urban boundary layer (free surface layer) is extended above the roof tops. The canopy sub-layer possesses its own flow characteristics that is mainly driven or determined by the interaction of the flow field above and the local effects such as building geometry, street patterns, urban traffic, natural topography and other local features like the presence of trees and landscaping structures. Generally, it is observed that wind speed in the canopy layer of the urban structure is seriously decreased compared to the undisturbed wind speed. Because of the reduction of the wind speed in urban street structure, there is significant limitation to the extent of which natural ventilation can be applied in dense urban environments[29]. The overall flow of wind near to the Earth's surface is often subjected to these boundary layer effects, referred to as the atmospheric boundary layer, by which the wind speeds are influenced by surface friction of the ground. The existing variations of different wind speed for heights on different urban terrains are illustrated below in fig. 13. Where the path of the wind flow is blocked by obstacles (such as trees and buildings), energy conversion occurs. The existent velocity pressure is thus converted to static pressure, in this way the windward side generates an overpressure of about 0.5–0.8 times the wind velocity, whereas an under-pressure results on the leeward side with about 0.3–0.4 times the wind velocity.

---

[28] *Geros, V., Santamouris, M., Tsangrassoulis, A., Guarracino, G. (1999). Experimental evaluation of night ventilation phenomena, Energy and Buildings*

[29] *Santamouris, M. (2001). Energy and climate in the urban environment. James and James Science Publishers, London*

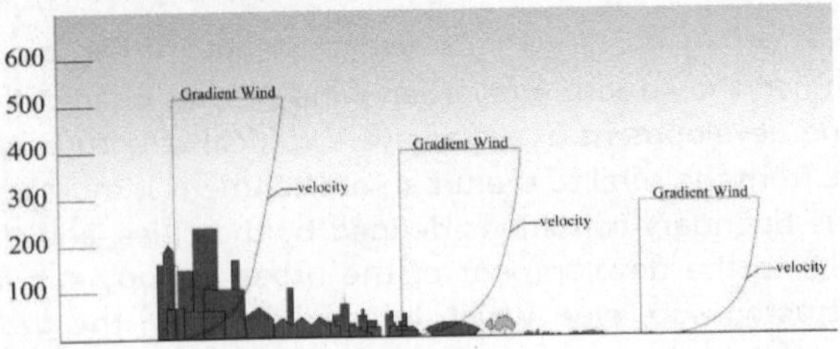

Fig. 13. Wind speed variation in urban environment, with height and terrain conditions[30]

Understanding the flow pattern and wind speed in urban street structure is therefore necessary for analyzing the natural ventilation potential for buildings in cities.

Due to impacts of urbanization, conditions like the heat island effects in cities coupled with global warming, has contributed to the increase of the peak cooling loads and electricity load necessary for building space cooling purposes, owing to this, it is significant to optimize building and urban space designs in terms of energy demands that con-siders local microclimates.

Many cities in Africa (especially sub-Saharan Africa) suffer from unplanned urban sprawls. These situations often arise from uncontrolled or poor developments. However, as the cities develop, they eventually become formalized and thus end up becoming major components of the urban development. Owing to the poorly planned conditions, build clusters raised in these areas face the huge challenges of poorly oriented plans, thereby raising houses that may face challenges of accessible wind flow for ventilation and hence, increasing reliance on electricity for cooling needs in high temperature seasons.

The ambient air temperature in the street levels in urban spaces often influences the energy load of the building environment and therefore determines the cooling potential of available natural and hybrid ventilation techniques. Concurrent thermal phenomena as well as recurrent vortices that often develop at street corners frequently dominate the

30 *Yang, Tong & Clements-Croome, Derek. (2018). Natural Ventilation in Built Environment.*

air flow patterns inside streets of settlements. Likewise, as solar heating of building exterior facade systems create surface temperature differences which also impacts on the air temperature conditions within the streets of settlements, such air temperature system eventually generate local upwind along warmer sides of building facade and form downward flows on the opposite side[31]. Furthermore, when wind speeds are small, thermal effects come into play within the surrounding environment, buildings tend to be warmer than the surrounding air and depending on the intensity of heating either the approach wind or the thermal convection dominate. The degree of the thermal change is dependent on many factors which may include the anthropogenic heat flux in the streets, sky view factor, and the albedo of the building surfaces. Also, the distribution of the air temperatures recorded at the top and at the different floors of buildings inside the streets of urban settlements indicated that temperatures above the street is often higher during the day time than the corresponding temperature inside the streets. Lower air temperatures within the streets were observed as the penetration of solar radiation is limited owing the depth of the street blocks. The lower temperature may be attributed to the mixing phenomena inside the street tunnels that tend to regulate the air temperature[32]. This shows that the air temperatures in the streets of urban building settlements are mostly governed by complex factors than normal ambient surface temperature despite the impacts of the local street configurations. When the airflow on an envelope surrounding the building is stressed, three main zones of disturbance are generated namely;

- Isolated rough flow,
- Wake interference flow
- Skimming flow.

Eddy vortex are formed in front due to flow down the windward face, while behind is generated a lee eddy which is

---

[31] *Georgakis, C., Santamouris, M. (2006). Experimental investigation of air flow and temperature distribution in deep urban canyons for natural ventilation purposes. Energy and Buildings*

[32] *Sakakibara, Y. (1996). A numerical study of the effect of urban geometry upon the surface energy budget. Atmospheric Environment*

drawn into the cavity of low pressure due to flow separation from the sharp edges of the building top and sides as shown in Fig. 14. These levels of airflow systems are necessary in the assessment of wind driven ventilation in urban buildings.

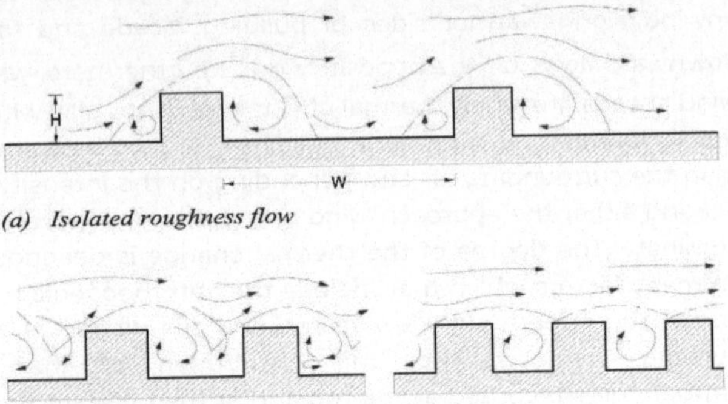

(a) Isolated roughness flow

(b) Wake interference flow *(c) Skimming flow*

Fig. 14. Representation of the three airflow systems in the urban boundary layer[33]

(a) is referred to as the isolated roughness flow regime,

(b) is referred to as the wake interference flow regime,

(c) referred to as the skimming flow regime.

---

[33] *Oke TR, Boundary Layer Climates. Routledge; 1987.*

CHAPTER SIX

Airflow Regimes

Studies of airflow regimes in built up settlements regard the geometry of the elements that form the urban fabric (relative to the height-to-width ratio) of buildings and the development orientations as the most significant parameter that influence urban microclimatic changes[34]. Such parameters as urban geometry are directly responsible for the quality of airflow at street levels and solar access, which in turn affects the quality of natural ventilation. Since air movement in urban settlements are dependent on interaction between impending winds within the built environment, so also does the pattern of existing wind alter in pattern when it flows through buildings. This impending airflow over urban settlements is subdivided into two layers that occur at the urban canopy level and the urban boundary level. With the impacts of the barriers (elements of urban geometry), airflow within the canopy urban level is often obstructed more than that of urban boundary level, resulting in wind driven air circulation used for natural ventilation in buildings. The entire circulation feature is regarded as the Airflow Regime. The three major flow regimes that exist in the Urban boundary layer is shown in fig. 15.

---

[34] *Todhunter PE. Microclimatic Variations Attributable to Urban Canyon Asymmetry and Orientation. Physics and Geography*

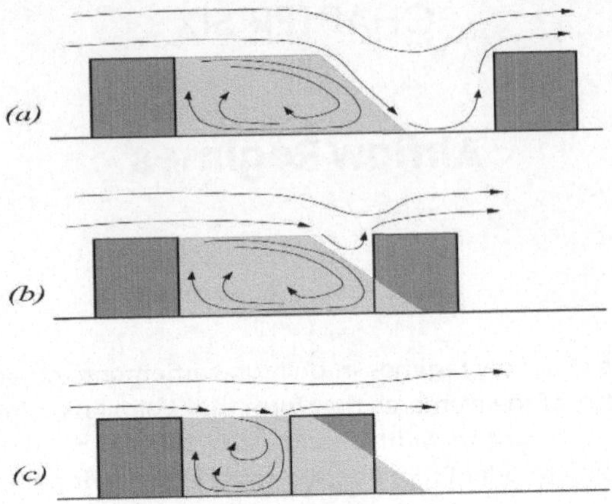

Fig. 15. Scheme of the three flow regimes in the urban boundary layer (Oke, 1987)

From Figure 15 above:

(a) is Lr < W (referred to as the isolated roughness flow regime);

(b) is Lr/2 < W < Lr (referred to as the wake interference flow regime);

(c) is W < Lr/2 (referred to as the skimming flow regime).[35]

The potentials of the wind circulation within urban developments are dependent majorly on the orientation and geometry (H/W and L/W) of built-up elements. Study on Airflow Regimes identified two regions which generate the wind patterns required for natural ventilation. The development of wind flows within these regions is likewise influenced by geometry and the fluxes scale differently owing to building canopies and site conditions. These two regions are:

- The Ventilated Region
- The Recirculation Region

---

[35] *Oke TR, Boundary Layer Climates. Routledge, 1987.*

CHAPTER SEVEN

Impacts of Airflow Regimes on Buildings

Two primary significance of controlled air flow in and around buildings are:

- Firstly is to meet the fresh air needs of building occupants,
- Secondly, to regulate indoor air temperature, humidity and pollutant concentrations. This also involves attention to the indoor environment, ensuring that the indoor conditions are thermally comfortable, with good air quality and moisture/temperature conditions do not promote condensation as well as fungal or bacterial mold growths etc.

It is established that majority of buildings are made up of individual component. Some of these components contain flaws, which occur during installations, particularly around openings in the envelope, such as around doors and windows, at junctions between wall, floor and roof and at sites of penetration for ducts, cables and piping. These flaws lead to air infiltration through holes, pores, cracks and fissures in the envelope. Buildings flawed envelope closures have relatively high infiltration rates and this eventually causes a thermal energy consequence leading to poor indoor thermal comforts and higher energy consumptions.

Since the features of envelops play major roles in the quality of airflow around and within buildings, it is important to understand the behaviors of wind flow around buildings in order to educate designers of buildings and Urban environments on the sustainable options for improving natural ventilation friendly environments.

One of the major considerations in evaluating Urban wind quality in the built environment is the Urban wind resource, which ultimately dependents on the urban micro-climate. The structure of the urban wind profile is composed mainly of the Urban Boundary Layer and the Urban Canopy Layer (as seen in figure 18). The three major scenarios that underscore the airflow behaviors on buildings in the Urban environment include:

1. Airflow behaviors above building canopy
2. Airflow behaviors within building canopy
3. Airflow behaviors below the building canopy

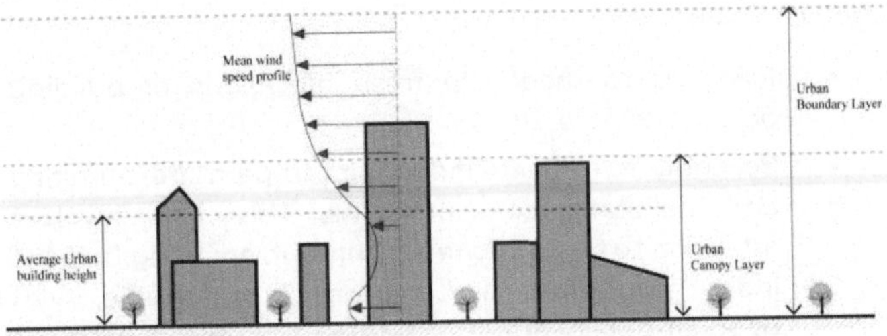

Fig. 18. Illustration of the Urban wind profile showing the positions of the Urban Canopy Layer and the Urban Boundary Layer[40]

Airflow Behaviors above Building Canopy

Generally, wind flow is instigated by the movement of air particles in the atmosphere owing to differences in pressure distribution in addition to the following:

i. Diurnal temperature variations,
ii. Seasonal effects, and
iii. Coriolis force exerted by the rotation of Earth.

Following above considerations, the characteristics and direction of wind flow is estimable from simple observation of the pressure and temperature distribution of air particles in Urban environments. This rational is only valid for the forecasting of airflow in higher altitudes and may not be

[40] *Ng, E., et. al (2011). Improving the wind environment in high-density cities by understanding urban morphology and surface roughness: A study in Hong Kong. Landscaping and Urban Planning*

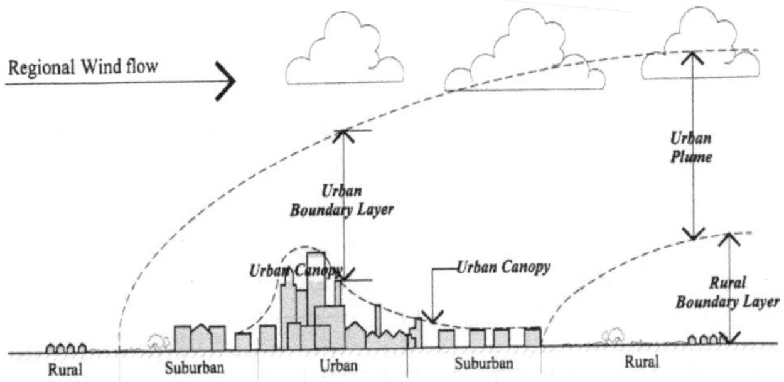

Fig. 17. Illustration of the urban atmosphere showing the two layer classification of Urban variation[39]

[39] *OKE, T.R., 1976, Inadvertent modification of the city atmosphere and the prospects for planned urban climates. Procedure Sympozium Meteorology Related to Urban and Regional Land-Use Planning, Asheville, N.C., World Meteorology Organization, Geneva*

The more active system termed the Ventilated Region, forms when the space within the courtyard cavity is sufficiently wide and this region provides the unobstructed flow for natural wind-driven ventilation. Conversely, a Recirculation Region develops at the near wake of buildings and in the case of courtyard buildings (dependent on building orientation), this system provides a weaker reused airflow system.

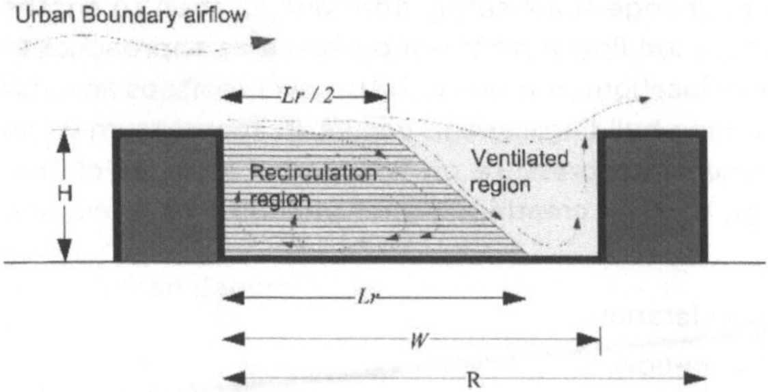

Fig. 16. Cross-section showing the Ventilated and Recirculation regions as they occur in Urban streets environments (Oke, 1987).

From fig. 16 above, the length of the recirculation region *(Lr)* is relative to the building height. According to studies, value of the ratio of the length of recirculation region and building height *(Lr/H)* depends on the turbulence levels in the urban boundary layer and the building geometry. Wind driven ventilation is achieved when the high speed flow from above the building roof level in the ventilated region is conveyed to ground level, thereby resulting in the development of internal bound flow layers along the surfaces. Equally, the continuous flow of wind slows within the recirculation region mostly due to the presence of milder moving air induced by the obstructive sides of the building.

Layers of Airflow Regimes in Built Environments

Rapid urbanization is among the numerous factors that influence and contribute to the elevation of air temperatures within the urban fabric. The results of these elevated temperatures and climate variance is responsible for the urban human discomforts. The adoption of natural ventilation in these built environments is one of the measures that may aid in limiting this effect of discomfort as effectively precise for calculating flow in lower altitudes. Other features and factors like surface roughness and topography of terrains that often dictate the amount of frictional force and shear stress exerted on the moving air particles may also significantly impact on the wind flow conditions within the low-altitude atmosphere[41].

A typical urban environment is composed of buildings, trees, poles, towers, etc., with diverse geometries along with other constant dynamic elements such as vehicles and human movements. These built structures and dynamic features constitute obstacles to the flow of wind in the urban environment. These obstacles increases the surface roughness and the complexity of the terrain, hence over the surface of an urban landscape, a boundary layer is formed, which comprise of the Urban Boundary Layer (also known as the atmospheric boundary layer) This Urban Boundary Layer is the lowest part of Earth's atmosphere and typically extends from the surface of Earth to a distance where the wind is no longer influenced by surface roughness. Within this Urban Boundary region exists three major sub-layers with corresponding velocity conditions[42]. Fig. 19 presents the layers of wind behavior over Urban environments.

Fig. 19. Illustration of the Urban wind flow and Urban Boundary Sub-Layers[43]

[41] *Nithya, D. S et.al. (2024). Review of Wind Flow Modelling in Urban Environments to Support the Development of Urban Air Mobility. Drones*

[42] *Grimmond, C.S.B., Oke, T.R. (2002). Turbulent heat fluxes in urban areas: Observations and a local-scale urban meteorological parameterization scheme (LUMPS). Journal of Applied Meteorology. Climatology.*

[43] *Grimmond, C.S.B., Oke, T.R. (2002). Turbulent heat fluxes in urban areas: Observations and a local-scale urban meteorological parameterization scheme (LUMPS). Journal of Applied Meteorology. Climatology.*

Airflow Behaviors within Building Canopy

Within the building canopy (Urban Canopy Layer), wind flow conditions are influenced by the immediate surroundings of the urban environment. The general expanse of the Urban Canopy Layer reaches to the roof level of tallest buildings in an urban environment. Within this layer also lies the roughness sub-layer, which stretches a few meters beyond the Urban Canopy Layer. The wind flow conditions within the roughness sub-layer are often in the process of adapting to the obstacles in the environment. Finally, the Inertial sub-layer appears beyond the roughness sub-layer leading to the top of the Urban Boundary Layer, where the wind flow conditions can be said to have fully adapted to the surface beneath it. Studies identifies wind flow velocity within these layers as always minimal as it gets nearer the surface of the ground due to surface roughness compared to the flow velocity at the top of the Urban Boundary Layer where the wind speeds are seen as identical to the *geostrophic* (Coriolis) wind speed[44]. Additionally, wind flow within the Urban Boundary Layer is always turbulent owing to the prevalent distortions in wind flow and surface temperature variations.

Airflow Behaviors below the Building Canopy

It has been noted that wind flow in urban environments are complex and turbulent in nature, as many obstacles with various geometrical forms and sizes impact on its behavior. The action of a steady flow of air particles directed on a building surface with normal conditions is expected to be redirected to varying directions, resulting in different flow patterns and forming other active sub-regions around the building. The patterns created emanate from the inactivity points and the flow is then deflected towards the lower pressure regions of the building like the sides, the top or the downward surfaces of the windward region of the building. This phenomenon is mostly due to presence of high pressure stagnation points that are activated on the windward surface

[44] *Britter, R.E.; Hanna, S.R. (2003). Flow and Dispersion in Urban Areas. Annual Review of Fluid Mechanics*

of buildings. Dependent on building geometry, the position of this inertia (inactive) points (mostly at two-thirds of the building's facade), is often dependent on other conditions like the proportion of building itself, building surface's aspect ratio, the Urban boundary layer height setting and finally the roughness upwind on surface of the building[45]. As winds meet a building, the deflection (change in direction) that occurs produces such varying effects on the building that determines the behavior of the areas surrounding the building. As this occurs, the Windward side (comprising of exterior walls and any other vertical surfaces facing the wind perpendicular to its path), are subjected to inward pressures called positive pressures. Conversely, as wind flow continues around and above the building, it also transforms to outward pressures (called negative pressure) as shown in figure 20.

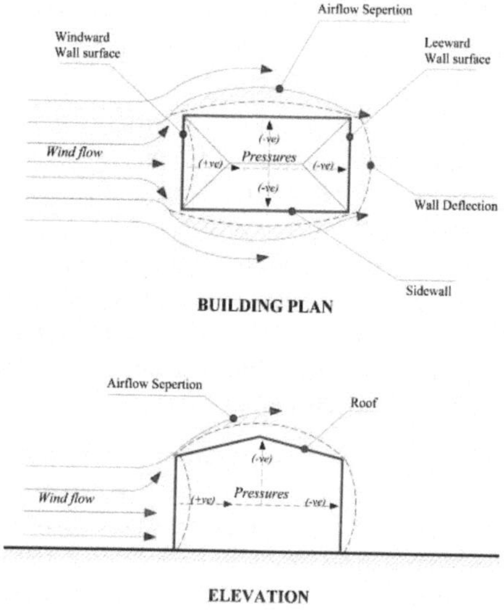

Fig. 20. Illustration of the behavior of Wind flow around a building (Peterka, et.al., 1985)

[45] *Peterka, J.A., Meroney, R.N., Kothari, K.M. (1985). Wind flow patterns around buildings. Journal of Wind Engineering and Industrial Aerodynamics*

These occurrences produce a form of suction effect otherwise referred to as outward (negative) pressures on the sidewalls as well as the Leeward wall and the roof. The wind pressures on the roof vary mainly based on the roof slopes and the location on the roof with the highest pressures occurring at the edges of the roof. The following effects are created along the roofing lines:

i. The roof surfaces on Windward area where the roof is a flat or shallow sloped type, will be subjected to a negative (outward) pressures.
ii. On the other hand, moderately sloped roofs (about 28°) are generally subjected to same negative pressure.
iii. In the case of very sloping roofs, the Windward side responds just as the adjoining walls, thereby sustaining the positive pressures, just as the sections of the roof on the Leeward side are subjected to dissimilar negative pressures.
iv. In the case of sloping roofs where wind direction occurs parallel to the ridge, the wind pressures would act just like an ordinary flat roof and is thus subjected to negative (outward) pressure.

Following the wind behaviors around a building, it is worthy of note that the Windward side of a building always carry the larger load (pressure) due to the wind pressure on it compared to the Leeward side, owing to the fact that it directly faces the oncoming wind.

In the case of the building surroundings within the building canopy, the behavior of the winds around the windward front of a building is such that a reverse flow occurs when the flow deflects downwards from the stagnation areas to the ground, therefore creating a detachment zone that forms vortexes near the building surface. Meanwhile, wind flow on the Leeward and sides of the building forms the *Wake* region. This Wake region forms vortexes which are characterized by a low velocity distribution and high turbulences (figure 21). Also within this Wake region, the wind flow then recirculates downstream making the wind flow highly turbulent[46]. In the case of wind flow around multiple building clusters with several building shapes, sizes and similar wind flow in and over the Urban Canopy Layer (with random roughness), studies identified that the behavior of wind flow is far more complex than the flow over a canopy region with isolated buildings of similar height[47].

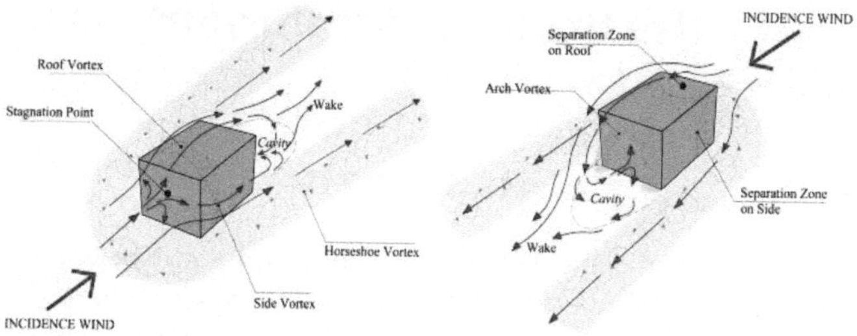

Fig. 21. Illustration of the typical behavior of wind flow around a single building with a perpendicular inflow/incident angle [48]

[46] *Hunt, J. (1975). Fundamental studies of wind flow near buildings. In Models and Systems in Architecture and Building—LUBFS; Conference Proceedings No. 2; Construction Press: Berlin, Germany.*

[47] *Xie, Z.T., Coceal, O., Castro, I.P. (2008). Large-Eddy Simulation of flows over Random Urban-like Obstacles. Bound.-Layer Meteorology*

[48] *Pourteimouri, P. et.al (2022). Numerical Study on the Impact of Building Dimensions on Airflow Patterns and Bed Morphology around Buildings at the Beach. Journal of Marine Science and Engineering.*

CHAPTER EIGHT

Analysis of Wind flow Patterns for activating Natural Ventilation in the Urban environment

The process of achieving natural ventilation is subject to the micro climate conditions of an urban environment. As discussed earlier, attaining this passive process is dependent largely on the ambient wind flow, radiation and air temperatures of the urban location. Significantly too, the urban space geometry is the most relevant parameter responsible for the microclimate variation at the micro level. Following the studies in wind behaviors, it could be deduced that when winds blow in the Urban environment, the following effects can be created (Fig. 22)[49]:

- Downward-deflection effects
- Upward deflection effect
- Side-ways-deflection effect
- Downwind eddy effect
- Counter-current effect
- Street canyon effect
- Venturi effect
- Wind tunnel effect

[49] *Aiyer, P. (2021). Wind Patterns in Built Environment. International Journal of Research and Analytical Reviews.*

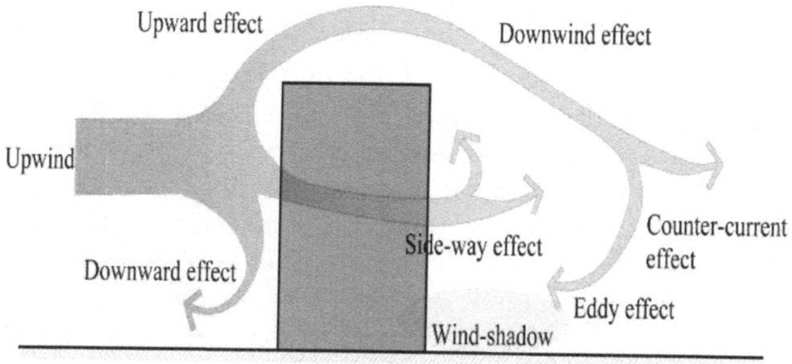

Fig. 22. Illustration of the effects created when winds meet buildings and other obstacles in the Urban environment[50]

i. Downward-deflection effects:

As heavy winds run into high rise buildings at greater heights, these winds are deflected downwards towards the streets and other open areas causing turbulence as they are caused to flow in different directions (see fig. 22).

ii. Upward deflection effect:

Upward deflections are created as wind moves with a force onto vertical structures. These structures (buildings) obstruct the flow, thus dispersing the wind into different directions with minimum resistance, both vertically and across the sides of the structure. Hence, the upward effect is that which deflects the wind upward after the obstruction happens, with no change taking place at the ground level (see fig. 22).

iii. Side-ways-deflection effect:

This effect is expected when windward surface of a building is either circular in shape or the wind speed is less on a more horizontal direction. In this case, the wind deflection goes sideways without affecting the airflow at the lower ground level (as seen in fig. 22).

---

[50] *Aiyer, P. (2021). Wind Patterns in Built Environment. International Journal of Research and Analytical Reviews*

iv. Downwind eddy effect:

At the point of vertical deflection and as the wind is moved sideways by the windward side of a building, the wind rebounds upwards to extended distances in the downwind direction, thus creating a calm wind shadow zone. The Eddy effects are the reversed currents that are created due to the pressure differences, resulting in either turbulence or light breezes (as seen in fig. 22).

v. Counter current effect:

In the case of winds that are caused to move in opposite direction to its flow, counter currents are created. The resultant gentle winds caused by the reverse deflection by obstacles (buildings) or by reverse flow of wind in the negative or zero pressure zones are thus referred to as the counter currents as seen in fig.22.

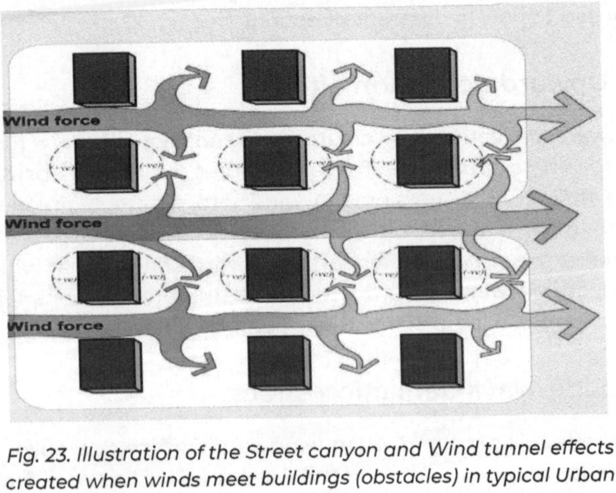

Fig. 23. Illustration of the Street canyon and Wind tunnel effects created when winds meet buildings (obstacles) in typical Urban streets[51].

[51] *Krishnan,A. Yannas,S. Baker,N. Szokolay,S. 2001. Climate Responsive Architecture. New Delhi: McGraw Hill Education*

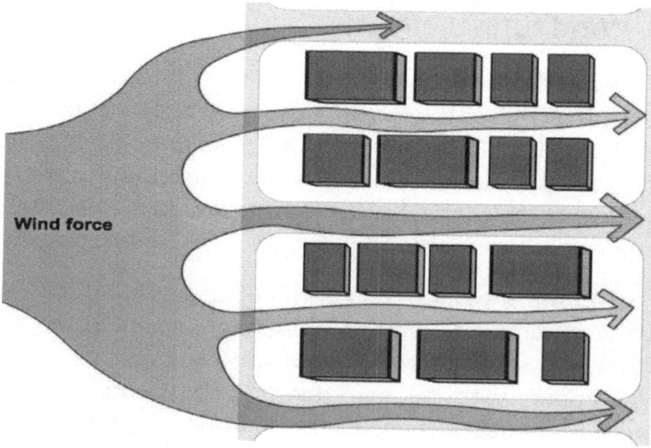

Fig. 24. Illustration of the Venturi effect created when winds meet buildings (obstacles) in typical Urban streets[52].

vi. Street canyon effect:

Cities in Urban environments comprise of buildings arranged with openings created for pedestrian/vehicular access otherwise referred to as streets. These streets form a structure of tunnels across the city's fabric and become tunnels for wind flow (fig. 23). As winds move across these canyons in perpendicular direction to the street, the resultant effect are pockets of turbulence between buildings. However in the case the wind comes at an angle to the street, the resultant effect then will be both street canyon and tunnel effects are observed. This effect is highly dependent on the street size and proportions of canyon.

vii. Venturi effect:

As stated in the Venturi principle (Giovanni Venturi, 1797), decrease in pressure occurs with increase in velocity. Thus the speed of air increases (without change in rate of flow and volume) as it moves through narrow spaces from wider ones (fig. 24). Winds flowing into Urban cities are often restrained by line of buildings in forms of tunnels, thereby increasing the wind speed significantly as it cuts through these tunnels.

---

[52] *Aiyer, P. (2021). Wind Patterns in Built Environment. International Journal of Research and Analytical Reviews*

viii. Wind tunnel effect:

As expected owing to the tunnel characteristics of Urban streets, the smaller the space between two rows of buildings would result in lower air pressure. Therefore having been deflected by buildings on either side of the narrow street, the movements of winds are caused to go faster in circling motion while passing through (fig. 23).

As seen from the discussion of the effects of wind flow into Urban environments, winds tend to interact with buildings in complex ways, resulting in various wind flow patterns. Urban environment with higher density constructions would impose more obstructions to the wind flow thereby slowing down the wind and deflecting it to different directions, resulting in complex behaviors. Similarly, through the discussion of urban air flow conditions, it could be deduced that the structure of wind fields can be characterized through the creation of vertical structure which occurs in two layers. Studies has identified that when the air coming from the rural environments flows to the urban areas, it naturally adopts the new boundary conditions created by the city's structure, thus resulting in the formation of the obstructed urban canopy sub layer, which is extended from the ground surface up to the buildings height, while the urban boundary free surface layer extends over the roof-tops. This urban canopy sub layer possesses specific flow properties which are defined by the interface between the air flow above the roof tops and the prevailing local conditions within the environment like the shape of buildings, street patterns, urban vehicular circulation, natural topography and the surrounding local landscaping structures like trees and shrubs etc[53].

Similarly, wind speed in the canopy layer of the urban structure is generally decreased compared to the undisturbed normal wind speeds. Because of the reduction of the wind speed in urban streets spaces, there is significant

---

[53] *Oke TR. Street design and urban canopy layer climate. Energy and Buildings*

limitation to the extent natural ventilation can be applied in dense urban environments[54].

Designers of urban spaces are mostly unfamiliar with the thermal comfort problems arising from the relationship between urban forms and local climatological conditions. Research identified that owing to these problems, many urban settlements do not respond to contextual climatic conditions. The common practice of using universal meteorological data gotten from stations that are located in unobstructed areas of the city like the airports or isolated fields may result in inaccuracies regarding the air flow in typical urban buildings as the microclimatic conditions play major roles in defining the eventual conditions of any given urban location.

It should be noteworthy that recent urbanization trends calls for sustainable ideas tailored towards improving thermal comfort. Understanding the flow pattern and wind speed in urban environments is therefore necessary for analyzing the natural ventilation potential of urban buildings, especially in the tropics which is characterized by high temperatures and humidity. The impacts of urbanization, the city heat island effects coupled with global warming contribute to the increase of the peak cooling loads and electricity consumption for space cooling purposes, owing to this and more, it is vital to optimize building and city designs in terms of energy demands that considers local urban microclimates.

This section of this book looks at the impacts of the elements that make up the Urban fabrics on wind flow patterns for passive ventilation purposes. It comprises of a review of analyses of the impacts of Building geometry as well as the impacts of other urban landscaping elements, on wind flow patterns. It also reviewed the impacts of Urban street configurations on wind flow for natural ventilation in typical Nigerian city environments.

---

[54] *Santamouris M. Energy and climate in the urban environment. London: James and James Science Publishers*

Analysis of Effects of Building Geometry on Wind flow Patterns

Owing to the natural phenomena of air movements, buildings and other manmade structures in the urban environment impede on the natural wind flows, thereby affecting its natural paths, velocity and directions. Likewise, the roads we create, carriageways and streets all turn into tunnels and canyons for wind flow, thereby causing disruption in the wind patterns. Architects and building designers face difficult decisions in identifying actual wind paths and direction in order to produce effective building orientations and layout designs that would capitalize on the wind driven natural ventilation. This demands that the wind flow conditions are studied to the minutest microclimates for building designers who intend to consider natural ventilation for thermal comfort and indoor cooling options. With the understanding of wind flow behaviors, designers can successfully produce buildings which provide options and elements that may be intentionally planned and installed to aid in projecting wind pattern and improve natural ventilation.

To throw more light into the analysis of the impacts building forms have on wind flow patterns, this section observes the activities of wind flow for natural ventilation in real case scenarios. A typical case is taken from the analysis of wind flow on a significant building type well known for its passive design performance in tropical Nigerian environments. The typologies of courtyard buildings are studied using parameters that identify with natural ventilation principles. These parameters, which include the measurements of airflow recirculation and ventilation regions affected by height and width ratios, are believed to have major effects on the airflow regimes experienced in urban environments.

Nigeria lies above the equator between latitudes 4° and 14°N, and longitudes 2° and 15°E making it a typical tropical location. Climatically, it falls within the area labeled as warm humid region. Its warm and humid climate, just like those of other sub-Saharan countries within the West African region, is influenced by two major factors[55], which are:

- Daily heating and cooling of the Sahara Desert land area; and
- Heating and cooling of the Atlantic Ocean.

These two phenomena as stated above create the distinctive dry and rainy seasons. It also creates the distinctive wind patterns which ideally are characterized into the North-Easterly (Harmattan) winds and the South-Westerly (Monsoon) winds as seen in fig. 25.

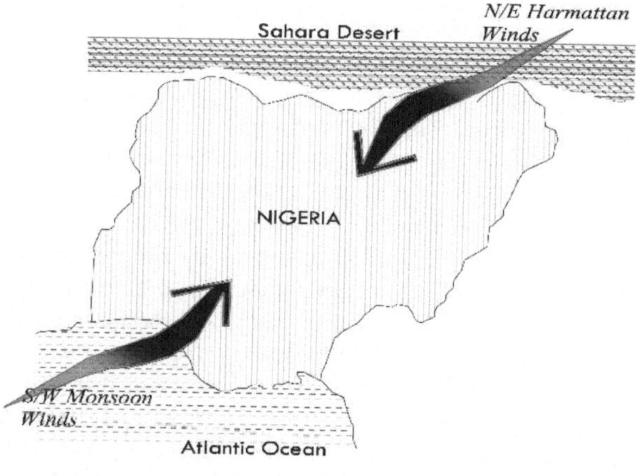

Fig. 25. Map of Nigeria showing the major seasonal wind patterns

Studies on thermal comfort, suggests that the effects of air temperature, humidity and air movement in hot climates can be used to determine the human thermal responses. In hot climates, the significance of air movement and relative humidity are mostly vital where the heat lost via evaporation often prevails. Research in tropical areas identifies the comfort limits for naturally ventilated spaces to range between 26°C to 31°C, with wind speeds of 1.0 m/s[56]. More so, within these hot humid (tropical) climates, high wind-speeds are often required to maintain thermal comfort ranges at

---

[55] *Ojo, O. (1967). The climates of West Africa. London: Heinemann.*

[56] *Boutet, T.S. (1987). Air Movement; A Manual for Architects Controlling and Builders. Mc-Graw Hill: New York*

higher temperatures. These impacts of natural ventilation mentioned above can be induced by wind pressure, temperature difference/Stack-effect or humidity difference. These pressure differences that results from wind flow is generated by temperature variations. Research on airflow regimes identifies that incident flow conditions, as well as the urban environment and building geometry can influence wind driven pressure distribution and wind direction. The relationship between building design and application of wind driven natural ventilation may thus lie on the performance of indoor air by the air pressure applied onto a building façade and the varying conditions of the outdoor air movement.

In Nigeria, most designers of public and residential buildings make conscious efforts to integrate concepts of natural ventilation in their designs. It is common believe that the courtyard style of building enables the capturing of natural wind flow for the purposes of passive ventilation and this has made the courtyard building typology very popular in this regards. However, very little is known by both students of architecture and building designers of the impacts the courtyard style building geometry has on wind flow patterns, which would indicate its practicability for natural ventilation application. A study of the effects of wind flow on the geometry of this building type (both on building surfaces and cavity spaces) is essential in understanding the actual effectiveness of this typology in application. Building features and geometry are among the key parameters that outline a courtyard's potential in trapping wind for purposes of natural ventilation, therefore it is essential to identify the relationship between courtyard proportions and the airflow patterns that may develop within them. By understanding the behaviors and patterns of airflow, designers will be equipped with ideas necessary for the composition of building forms that will enable the achievement of courtyard buildings that are more naturally ventilated.

In Nigeria, different configurations of courtyards are adopted in different locations to serve a variety of functions. However, the primary function is that of passive cooling through natural ventilation. Studies of existing courtyard configuration all across Nigeria identified a variety of compositions which could be categorized as vernacular, colonial, post-colonial and modern as shown in Table 1. This concept of architecture is seen in various building types ranging from residential buildings to administrative, institutional, health, educational, and multi-functional buildings.

Table 1. Classification of courtyard building typology found in Nigeria[57]

Vernacular	Colonial	Post-colonial	Modern

Table 2. Various functions of courtyard buildings in Nigeria

Institutional	Health	Administrative	Residential

For the purpose of this analysis, different forms (outlines) of courtyards adopted in various building types all across Nigeria were collected and classified based on their geometrical characteristics. These geometrical forms are the major containers for wind flow collections in courtyards. The classification is shown in table 3, indicating the ranges of the horizontal (W,L) and vertical (H) proportions.

Table 3. Various configurations of courtyards applied in buildings in Nigeria

TYPE	BUILDING TYPE	WIDTH RANGE	HEIGHT RANGE	H/W RATIO	CONFI. PLAN	CONFIG. SECTION	CAVITY GEOMETRY
NARROW		<50m	<12m	0.24	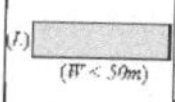		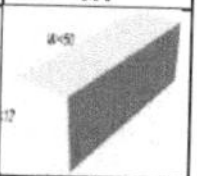
WIDE		<30m	<6m	0.20	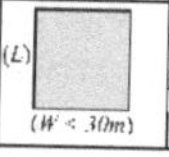	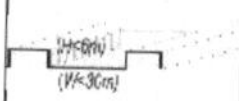	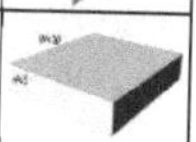

---

[57] *Akubue, Jideofor & Adesina, Damilola. (2023). The Effect of Courtyard Geometry on Airflow Regimes for Ventilation in Tropical Nigerian Environment. Journal of Architectural Environment & Structural Engineering Research.*

HIGHRISE		<50	<36m	0.72	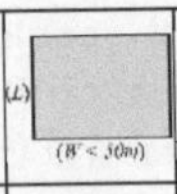	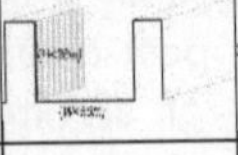	
LOW RISE		<10m	<4m	0.40	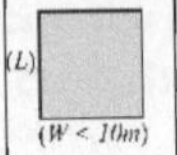	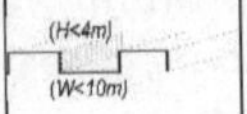	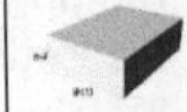
DEEP		<28m	<35m	1.25	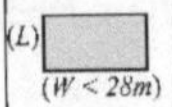	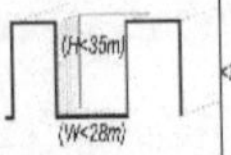	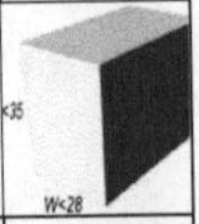
MODERATE		<50m	<18m	0.36	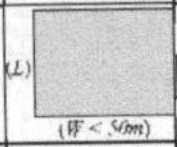	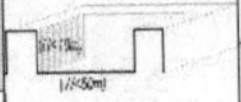	
TIGHT		<6m	<8m	1.30	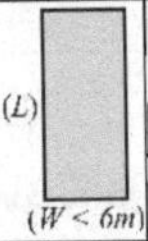		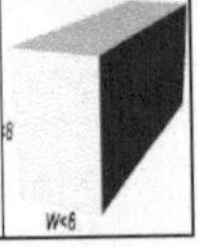

As identified in the discussion of "Airflow Regimes" in the introductory section, the three airflow regimes responsible for the urban air circulation system develops two distinct regions termed Ventilated and Recirculation regions. The more active system termed the Ventilated Region, forms when the space within the courtyard cavity is sufficiently wide and this region provides the unobstructed flow for natural wind-driven ventilation. Conversely, a Recirculation Region develops at the near wake of buildings and in the case of courtyard buildings (dependent on building orientation), this system provides a weaker reused airflow system. Taking into consideration the Length of Recirculation (*Lr*) and the ratio of the length of recirculation region and building height (*Lr/H*) as seen in Fig. 15 (AIRFLOW REGIMES), wind driven ventilation is achieved when the high speed flow from above the building roof level in the ventilated region is conveyed to ground level, thereby resulting in the development of internal bound flow layers along the surfaces. Equally, the continuous flow of wind slows within the recirculation region mostly due to the presence of milder moving air induced by the obstructive sides of the building.

Through this analysis, the likely performance of each courtyard configuration (as seen in Table 3) for effective natural ventilation can be predicted.

Through the calculation of the ratios of recirculation to building height in each of the configurations, the predicted performance of each building type based on the area of Ventilated regions formed in each courtyard cavity configuration is presented in Table 4.

Table 4. Predicted airflow performance of the different courtyard configurations

Courtyard type	Courtyard Configuration	Courtyard Ratio	Projected airflow systems	Resultant Airflow Regime
NARROW		0.80	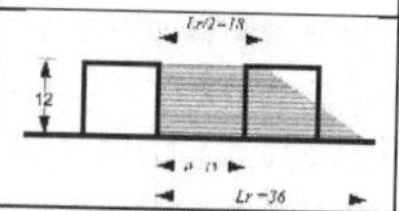	Skimming regime
WIDE	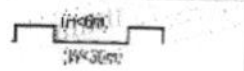	0.20	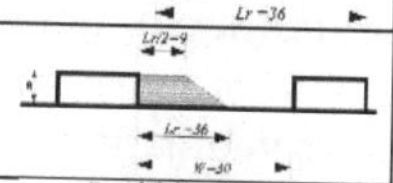	Isolated roughness regime
HIGHRISE	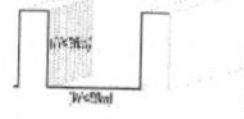	0.72	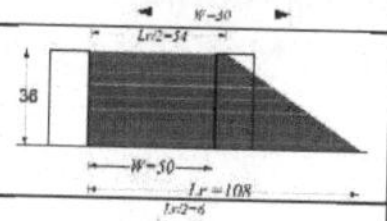	Skimming regime
LOW RISE	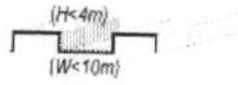	0.40	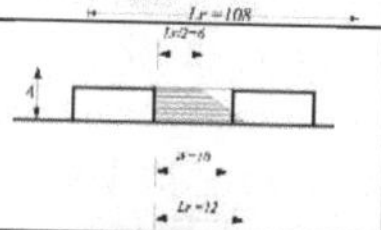	Wake interference regime
DEEP	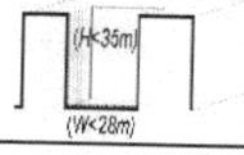	1.25	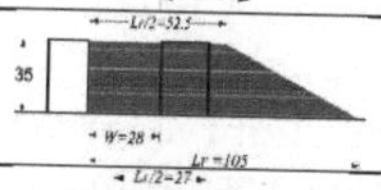	Skimming regime
MODERATE	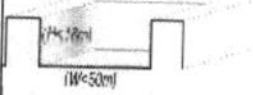	0.36	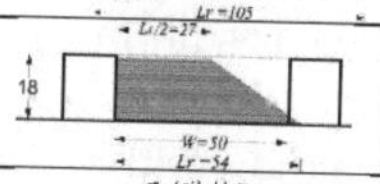	Wake interference regime
TIGHT (Width)	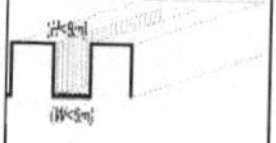	1.30	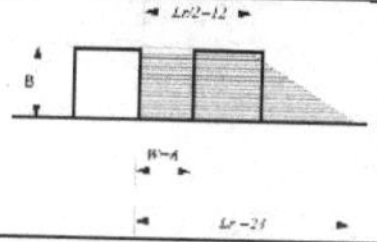	Skimming regime

Note: Courtyard cross-section showing the Ventilated and Recirculation regions

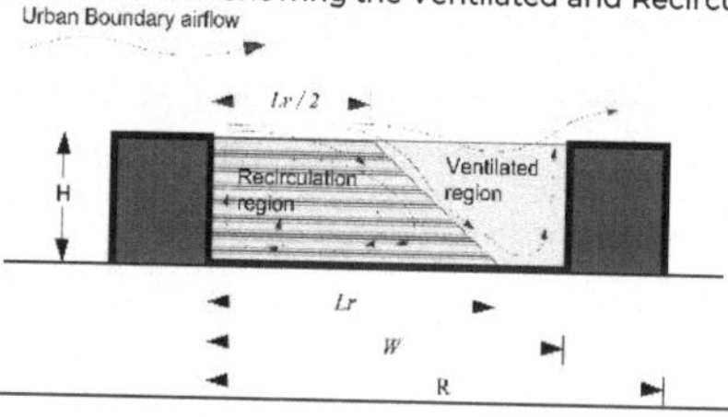

Following the analysis of the airflow systems in the different courtyard configurations as shown above, the potential

scenarios that may ensue relative to geometry of the courtyard building types can be identified. Let it be noted that Table 4 presented the flow regime schematics for the dissimilar H/W configurations while assuming the value of Lr/H=3. The objective of this is to graphically identify the extents of ventilated and recirculation areas accessible for natural ventilations as well as to ascertain the extents of obstruction produced by the building geometry itself on the airflow in this region.

From the analysis as shown in Table 4, it is evident that the Wide Courtyard typology with a courtyard (H/W) ratio of 0.2 is the only configuration that results in an isolated roughness flow. Isolated roughness regime occurs between well-spaced building units with reduced obstruction for urban boundary airflow that drives ventilation in buildings. On the other hand, the Low-rise and Moderate typologies with courtyard (H/W) ratios of 0.4 and 0.36 respectively present the wake interference flow with limited ventilated airflow regions to the windward face of the courtyards. This limited airflow from the urban boundary layer occurs due to obstruction caused by the higher building geometry, hence permitting ventilated airflow on partial areas of the internal courtyard façade. The other four types which include the Narrow, High-rise, Deep and Tight configurations with courtyard (H/W) ratios of 0.8, 0.72, 1.25 and 1.3 all result in Skimming flow. The Skimming flow regime as discussed earlier is totally dependent on airflow within the urban canopy layer owing to obtrusiveness from the geometry of the narrower courtyard widths relative to the building heights. Main features of skimming airflow are the slow wind speeds from the recirculation regions resulting in weaker velocity for driving natural ventilation in the buildings.

These three scenarios are summarized in Table 5, showing the courtyard (H/W) ratio with best potential for natural ventilation.

Table 5. Comparison of courtyard geometry (H/W ratio) and natural ventilation potentials [58]

	Courtyard (H/W) ratio	Extant airflow regime	Natural Ventilation potential
1.	0.1 – 0.2	Isolated roughness airflow regime	Good
2.	0.3 – 0.5	Wake interference airflow regime	Fair
3.	0.6 – 1.0 and above	Skimming airflow regime	Poor

It is worthy of note that the benefits of courtyard buildings cuts across introducing cross ventilation in room spaces and improvement of thermal comfort through air movement in hot tropical environments. Through analysis of different courtyard configurations stated above, the natural wind-driven ventilation prospects in each of the configurations admits that the wider-spaced design type is most suitable. This also signifies that the H/W ratio that falls between the ranges of 0.1 and 0.2 are most effective for introducing airflow with significant wind speeds for effective cross ventilation. Nevertheless, the effectiveness of employing natural ventilation in various locations against different courtyard configurations is also hugely dependent on the building orientation and planning. Students of architecture and building designers may adopt this analysis in determining the configurations to be applied during the design process, especially for buildings that intend to integrate natural ventilation for indoor space cooling.

Analysis of Impacts of Street Configuration on Airflow for Natural Ventilation

The typical urban fabric is made of physical features like buildings etc., arranged in patterns that are defined by the arrays of carriageways that spawn through them forming streets and corridors. The most significant physical elements that influence the behavior of wind flow in urban

---

[58] *Akubue J.A. (2019). Effects of Street Geometry on Airflow Regimes for Natural Ventilation in Three Different Street Configurations in Enugu City. IntechOpen. Different Strategies of Housing Design*

environments are thus identified to be the structure (buildings) and the streets (urban corridors). Having demonstrated a typical procedure for analysis of effects of wind flow on building geometry, it is therefore necessary to also demonstrate real case procedure for the analysis of the impacts street configurations have on wind flow patterns.

For this purpose, this section observes the activities of wind flow for natural ventilation in real case scenarios. A typical case is taken from the analysis of wind flow in tropical Nigerian urban environment. The significance of this analysis is to identify the primary influences, behaviors and parameters of flow within the urban streets relative to natural ventilation in regions in the country. These parameters, which include the measurements of airflow recirculation and ventilation regions affected by height and width ratios, are believed to have major effects on the airflow regimes experienced in urban streets.

Using Enugu metropolitan area as a case, the study utilized schematic analysis process to examine three major street configurations that stand out in the residential districts. This process forms a basis for the analysis of airflow regimes and its effects on natural ventilation. Using the parameters mentioned above, each of these street configurations in the Enugu residential district is analyzed in order to justify the effective street size for natural ventilation in this region. The analysis identified the possibility of wind flow reduction in urban streets, because of this reduction, there exists a significant limitation in the extents natural ventilation may be utilized in dense urban settings like the tropical settlements in Enugu city.

Enugu city (6.26° North and 7.29° East) is located in the Southeastern part of Nigeria. Its climate is classified as a tropical rain forest with a derived savannah. The city is set on an altitude of 304.7m above sea level, its climate is naturally humid and this humidity is regarded as oppressive between March and November. Its average maximum temperature stands at 34.9°C, with average lows of about 22.3°C and its annual mean temperature is about 26.7°C. Typical with the rest of West Africa, the rainy and dry seasons are the only weather periods in Enugu. The average hourly wind-speed in

Enugu area experiences significant seasonal variations over the course of a year. For instance, the windier part of the year lasts for 7 months (between March to September), with average wind speed of 5.5m/s. The windiest month of the year is usually August, with an average wind speed of 7.1m/s. Typical wind path comes relatively from the west for an average of 10 months (termed the monsoon season), ranging from January to November, with peak percentage of about 75% occurring in July. Also, the wind direction changes to northerly for 2 months (termed the Harmattan season) from the months of November to January, with 41% occurring in the peak month of January. Fig. 26, shows a typical climatic data of Enugu indicating the conditions discussed above. The outdoor wind data used in this study were obtained from the Nigerian Meteorological Agency (NIMET), at a height of 10m above ground level.

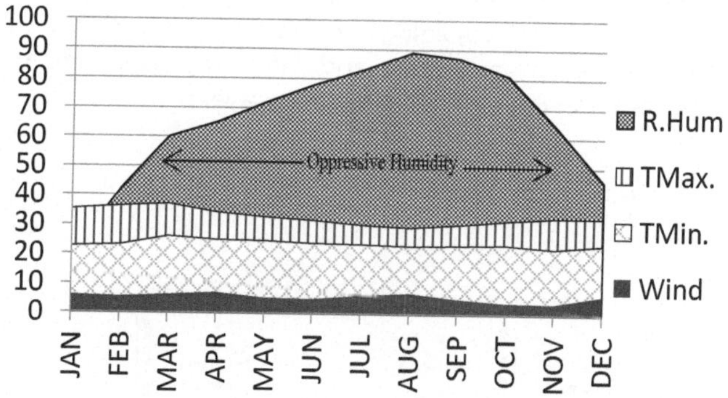

Fig. 26. Climatic Data of Enugu City (Source: NIMET)

Case 1. Analysis of Three Different Street Configurations:

The urban street has evolved from the original context as defined by the Latin word strata (meaning “a paved road”). In modern times, the function of the street in a built environment consists of more than the conventional roadway. Streets provide accessibility, connectivity, circulation, as well as infrastructural network for human settlements. As human activities grow, the function and configuration of a street system transforms especially in urban settlements. These transformations affect the micro climate environment within the street level as well as the universal urban environment. Urban climate is a critical factor which affects regional and global climates and consequently urban livability. Geometry and orientation of an urban street affect outdoor and indoor environments, solar access inside and outside the buildings, the permeability to airflow for urban ventilation, as well as the potential for cooling of the whole urban system[59]. A typical urban street is comprised of stationary and mobile elements which form barriers that collectively affect the flow of wind within the streets. Owing to the presence of these obstacles like buildings, trees and street furniture, the air flow through the canopy layer is frequently obstructed compared to the air flow within the boundary layers. These results in the gentler airflow observed within an urban canopy layer compared to the airflow within adjoining rural settings. Fig. 27, shows a typical urban street system and the barrier elements that affect airflow within the urban canopy layer.

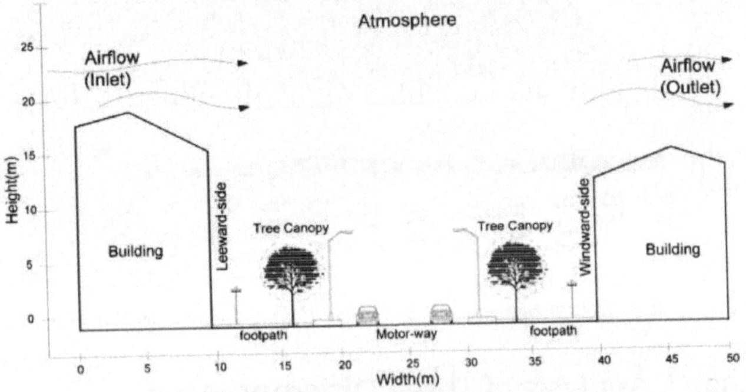

Fig. 27. Illustration of typical Urban Street showing the barrier elements that affect airflow within the urban canopy layer

Urban streets in tropical cities like the city of Enugu are structured to serve the basic conventional functions of

---

[59] *Ali-Toudert F, Mayer H. Numerical study on the effects of aspect ratio and orientation of an urban street canyon on outdoor thermal comfort in hot and dry climate. Buildings and Environment.*

accessibility and circulation. The configuration of Enugu streets is defined by urban planning classifications. The major classifications are the high density, medium density and low density. The low and medium density areas consist mainly of single villa residential housing units, while the high density areas consist of residential apartments. Three major residential districts stand out in these configurations (shown in fig. 28), these districts include;

i. Achara layout
ii. Independence layout
iii. GRA district

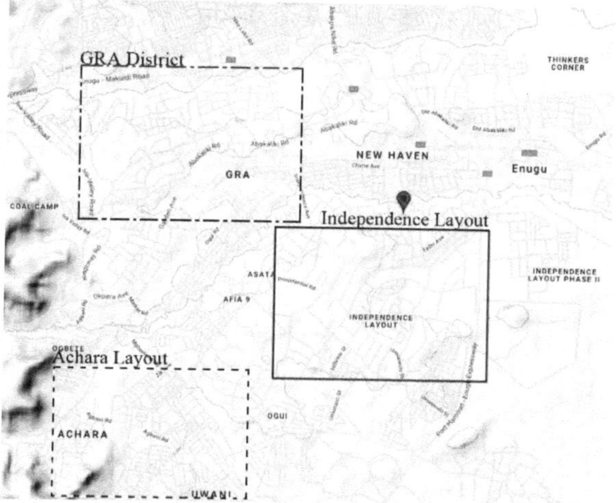

Fig. 28. Google Map showing the three major residential districts in Enugu city, representing the high, medium and low density districts

Table 6, shows the configurations of the streets in Enugu relative to Height and Width ratios. Unlike the typical European streets characterized by activities like walkways, plazas, pavement cafes and other lifestyle facilities which define the size of the street outline (urban canyons), the sizes of the typical street canyon in Enugu city is more compact. This is condensed further by the presence of property boundary walls which is used mainly as solutions to the poor urban security.

Table 6. Configuration of urban streets in Enugu city

Configuration	H/W Ratio	Planning Classification	Street Image	Aerial Image
	0.44	<u>GRA District</u> High rent residential district (bungalows and single home units)		
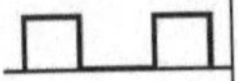	0.68	<u>Independence Layout</u> High rent residential district (2 floor homes made up of duplex and town houses)		
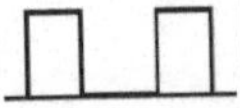	0.94	<u>Achara Layout</u> Medium rent residential district (apartments and blocks of flats)		

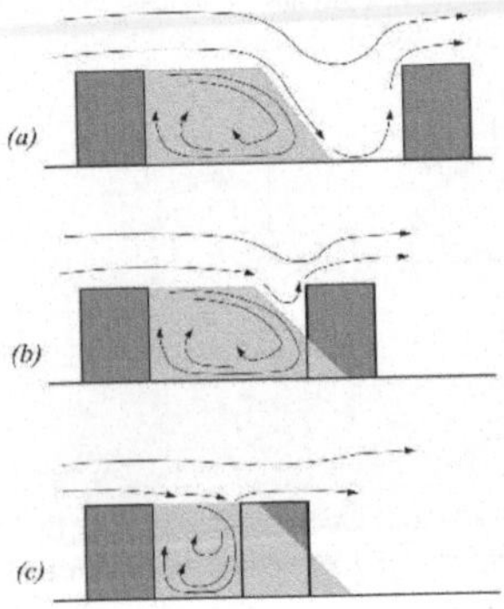

From the Illustration of the three major Air Flow Regimes (Fig. 15):

(a) is Lr < W (referred to as the isolated roughness flow regime);

(b) is Lr/2 < W < Lr (referred to as the wake interference flow regime);

(c) is W < Lr/2 (referred to as the skimming flow regime)

Following the equations above, it suggests that Lr/H should be between 2 and 3. In the case of wide street canyons, it suggests that Lr < W, so that H/W < 1/3, in which case, the recirculation region does not impinge on the downstream building as can be identifies in *(a)*. This it termed the "Isolated flow regime". In the case of intermediate (not so wide) street canyons, it suggests that Lr/2 < W < Lr, so that 1/3 < H/W < 2/3, in which case, the recirculation region begins to impinge on the downstream building as observed in *(b)*. This it termed the "Wake interference flow regime". Finally, in the case of narrow street canyons, it suggests that W < Lr/2, so that H/W > 2/3, in which case, the entire canyon volume is occupied by the recirculation regime as observed in *(c)*. This it termed the "Skimming flow regime".

Inside the ventilated region, high-speed air flowing from above the roof top level is conveyed to street ground level. This causes the development of internal boundary flow layers along the surface of the street level. As this occurs, the vertical field of the wind regulates to a layer of even equilibrium with the normal surface of the street. However, the airflow in the recirculation region is dragged by recurrent high momentum injections of wind burst consistently drawn off the upstream roof. The steady burst of airflow slows as it advances around the recirculation region owing to the occurrence of gentler moving air induced by the associated street barriers[60].

The analysis of the airflow patterns in Case 1, the residential streets in the tropical hot and humid Enugu city is presented using the flow regime analysis schematics for the different H/W configurations and assuming Lr/H=3. The objective is to identify the extent of obstruction caused by existing street geometry on the airflow for Natural Ventilation in this region using schematic diagrams.

The three different districts observed and shown in Table 6 above, have H/W ratios of 0.48, 0.68 and 0.94 respectively. For the 0.48 streets, the street width is approximately 16 meters and the average building heights is approximately 7 meters. In the case of the 0.68 streets, the street width is about 16 meters and the average building heights is roughly 11 meters. The 0.94 streets have street width of about 16 meters and the average building heights is approximately 15 meters. The schematics of the airflow regime for these streets are shown in Table 7.

---

[60] *Townsend A.A. The structure of turbulent shear flow; Boundary layers and wall jets. 2nd edition. London: Cambridge University Press; 1976*

Table 7. Schematic analysis of the airflow regime for the three main street configurations in Enugu city

<table>
<tr><th>Scheme</th><th>Result</th><th>Effective Street width</th><th>Resultant Airflow Regime</th></tr>
<tr><td>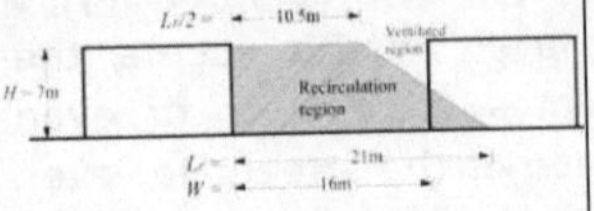

<table>
<tr><th>H/W</th><th>Lr / H</th><th>Lr</th><th>Lr / 2</th></tr>
<tr><td>0.44</td><td>3</td><td>21m</td><td>10.5 m</td></tr>
</table></td><td>Lr/2 < W < Lr
&
1/3 < H/W < 2/3</td><td>Intermediate-wide</td><td>Wake interference flow</td></tr>
<tr><td>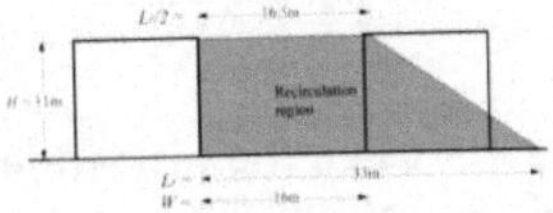
<table>
<tr><th>H/W</th><th>Lr / H</th><th>Lr</th><th>Lr / 2</th></tr>
<tr><td>0.68</td><td>3</td><td>33m</td><td>16.5 m</td></tr>
</table></td><td>W < Lr/2
&
H/W > 2/3</td><td>Narrow</td><td>Skimming flow</td></tr>
<tr><td>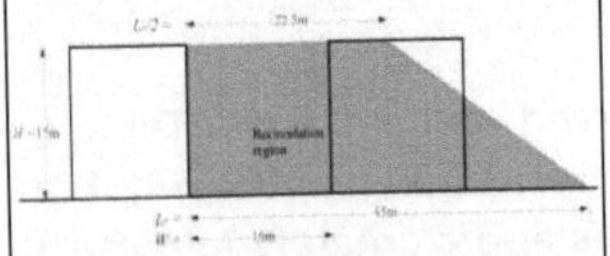
<table>
<tr><th>H/W</th><th>Lr / H</th><th>Lr</th><th>Lr / 2</th></tr>
<tr><td>0.94</td><td>3</td><td>45m</td><td>22.5 m</td></tr>
</table></td><td>W < Lr/2
&
H/W > 2/3</td><td>Narrow</td><td>Skimming flow</td></tr>
</table>

From this analysis, it could be noted that the isolated roughness regime, which is the airflow that takes place between well-spaced buildings is not readily present in the urban streets of Enugu. From the schematic analysis above, the intermediate-wide street category only showed the potential to allow the continuous airflow from the urban boundary layer to the buildings, due to the lower heights of the buildings. This means that the speed of the airflow from the urban boundary layer is not disturbed at the windward face of the buildings, therefore permitting ventilated airflow. However, this potential is diminished by the presence of continuous often solid concrete fence walls with average heights of about 2.4m (see Fig. 29) which is often employed as security measures in buildings in this city. The other two categories in the Skimming flow regime are entirely dependent on airflow at the urban canopy layer which is the skimming airflow. This is due to the narrower street widths resultant from the building heights. Characteristics of skimming airflow include slower wind speeds and re-circulated airflow which lacks the speed for adequate permeability of natural ventilation in the buildings within these streets. The use of concrete fence walls is also rampant in this category of streets in Enugu which impedes further on the speed of the skimming flow. It is noteworthy however, that the orientation of a street, building elements (barriers) and prevalent wind directions ultimately affects the permeability of airflow for urban ventilation.

Fig. 29. Varieties of concrete walls used as boundary/party fences in Enugu city

Following the schematic analysis, Table 8 presents the natural ventilation prospects in each of the three configurations shown above. With these scenarios, natural ventilation considerations could be carried along during the processes of building and urban street designs. However, the effectiveness of utilizing natural ventilation for different locations is improved through the orientation of streets and building planning.

Table 8. Natural Ventilation prospects for the three different street configurations

Configuration Schematic	Natural Ventilation Prospects
Configuration 1 (0.44): GRA District 	<u>Intermediate-wide street</u> • Allows for continuous airflow from the urban boundary layer to the buildings (ventilated region), due to the lower heights of the buildings. • Unobstructed wind flow from urban boundary layer at the windward face of the buildings • Tendency of faster airflow for Natural ventilation
Configuration 2 (0.68): Independence Layout 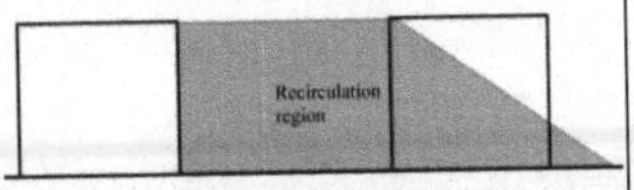	<u>Fine-Narrow street</u> • Allows airflow mostly at the urban canopy layer • Wind flow is skimming type as it is mostly re-circulated airflow. • Slower winds lack the speed for adequate permeability of Natural ventilation • Natural ventilation in buildings can be improved by applying openings on the un-shaded faces in the schematic (ventilated region)
Configuration 3 (0.94): Achara Layout 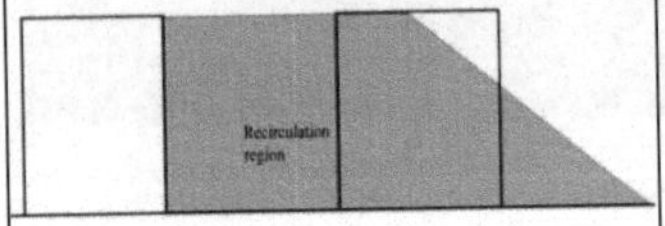	<u>Narrow street</u> • Allows airflow mostly at the urban canopy layer • Wind flow is skimming type as it is mostly re-circulated airflow. • Slower winds lack the speed for adequate permeability of Natural ventilation • Natural ventilation in buildings can be improved by staggering buildings to break the continuous street canyon structure

Through the analysis of the airflow regime for the 3 main street configurations in the tropical Enugu city, it could easily be concluded that the objective solution for passive ventilation is to design streets with wider widths (i.e. smaller H/W ratios). A wider street provides better mixing of air and consequently better airflow in the urban street canyon. But this can never be the case in most modern city developments owing to land use objectives. Consequently, the other option is the provision of adequate openings between streets and courts and adopting different building heights in the streets which could improve air exchange within the urban canopy layer.

Case 2. Analysis of Airflow in Typical High Density Street Configuration:

For this study, the typical '*Achara Layout*' (Medium rent) high density building cluster in Enugu city was selected. The surroundings of the measurement sites can be characterized as typical urban land-use, with residential housing. The street orientation has an effect on the street microclimate; hence the sites considered have an average of three storey buildings (4 floors) with a NE-SW street-axis orientation along an asphalt road.

The medium rise building heights in the location are averagely 16.5m on either side of its streets. The effective width of the streets is 18m, thus yielding a height to street width ratios (H/W) of about 0.92

The district considered for the study is characterized with medium high-rise buildings. Medium rise buildings are defined as buildings that fall within the range of 4 to 10 floor heights. The Achara layout district in Enugu city (indicated with the red circle in figure 2), is notable for its cluster of 4 to 5 storey high buildings, which are all apartment buildings or residential blocks of flats (fig. 30).

(a)

(b)

Fig. 30. (a) Aerial map of Enugu city, showing the study area in red circle, (b) typical street view of the Achara layout district

In Case 2 analysis, field investigation was conducted in order to ascertain the extent of the physical wind current available for driving ventilation within the streets of the selected district. To obtain the values necessary for this study, hand held equipment were utilized, which were operated by selected group of students placed systematically at significant points within the study area. By introducing students to the consequences of non-utilization of ambient microclimate conditions in building designs, the awareness and methods of recording and documenting microclimatic elements was created. The groups took turns to ensure proper records were taken. During the exercise, the study coordinator ensured synchronized timing for measurements, and collation of data collected from the participants. The required data taken during the study was the air temperature, wind speed (velocity), wind direction and air humidity values. The month of August was picked for this study, owing to its significance as the month with the highest annual wind speed in Enugu city. The values of the field measurements at each assigned point within the streets were used for comparative analysis to ascertain the significance of the available wind force within each street zone. The recorded wind speeds are identified as the contingent wind force (post obstruction) that these streets depend on for natural ventilation.

The field measurements were executed on selected locations for the assumed windiest month of the year (August). The point of this is to evaluate the wind flow and wind direction across these selected points for characteristic climate month and for 6 hours of the selected non-rainy day. Due to security preference the daylight hours of a working week day (8:00, 10:00, 12:00, 14:00, 16:00, 18:00) was used for the measurements. Wind velocity, air temperature and wind direction were measured using portable Digital Anemometer with sensor (LCD Digital CFM, CMM Airflow Gauge Meter, accuracy = + 3%+0.2m/s, with response time of 1 second) fixed on the selected points (at height of 3m) and recorded simultaneously at exactly these hours of the measuring day of the study. Fig. 31 shows the Google map of the study area as well as the designated points (points 1 to 18) where measurements were taken. Table 9 presents the summary of values recorded on the day of study.

For the purpose of this analysis, the streets in the study district were categorized according to the openness of its tunnels for receiving re-circulated wind flow as seen in figure 32. The following characteristics are identified:

- Arterial Street with multiple branches (intersections) as seen in streets A&B (fig. 33).
- Arterial Street with single branch as seen in streets C&F (fig. 33).
- Continuous Street with no branches or intersections as seen in streets D&E (fig. 33).

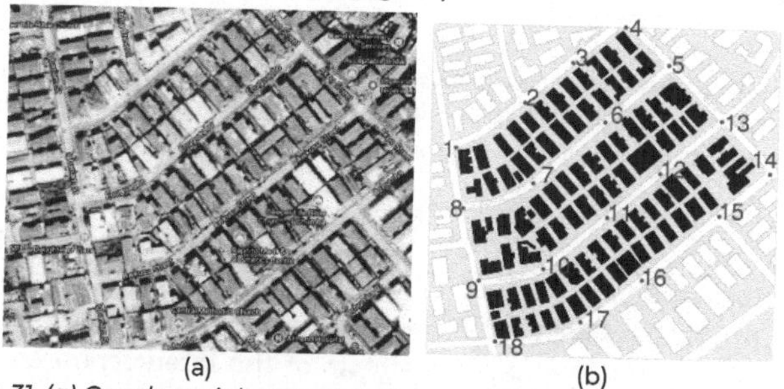

Fig. 31. (a) Google aerial map of study area, (b) Aerial map showing the designated points for field measurements

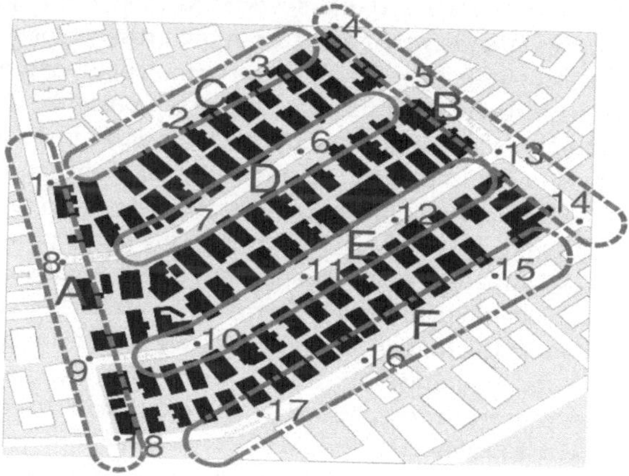

Fig. 32. The six streets (A,B,C,D,E,F) used for the study, showing the measuring points 1-18

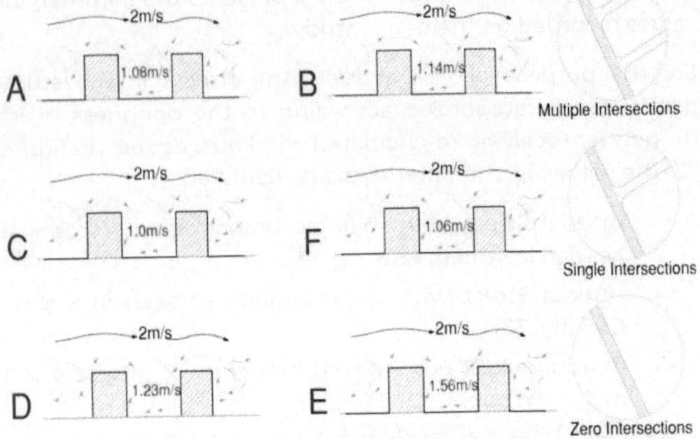

Fig. 33. Section of the streets showing the average monthly wind speed for the month under study (as 2m/s) and the average wind velocity obtained in each street (A,B,C,D,E,F) during the study day.

While streets A,B,C&F had a number of branches/ intersections across its canyon, only streets D&E provided tunnels that can enable continuous airflow across the street lengths (end to end). The effects of the intersection can be seen from the recorded wind velocity values in Table 9. The continuous streets (D&E) presented higher wind speeds than the streets with branched intersections as seen in the graph (fig. 34). This suggests that apart from the presence of street features like fence walls, trees cars etc., the speed of airflow within street tunnels are also affected by the presence of the intermittent inter-sections. Also, from the values in Table 9, whereby summarized in fig. 35, the seasonal monsoon wind system experienced during the season was seen to be consistent with a SW flow in the continuous street patterns. Apart from apparent obstructions from street features, this result significantly suggests that the SW pattern for street orientation in this region is most effective for driving wind currents for operative ventilation not only during the Monsoon season as seen above, but also for the Harmattan season with Northeasterly flow in the same street orientation.

Table 9. Showing the values of the measured points (H=humidity, V=air velocity, Dir=wind direction).

TIME	POINT 1			POINT 2			POINT 3			POINT 4		
	H	V	Dir	H	V	Dir	H	V	Dir	H	V	Dir
8:00 AM	80.7	1.3	SE	82.9	1.0	SE	80.0	1.8	SW	80.0	1.1	SE
10:00 AM	73.4	1.1	NE	72.3	0.6	NE	71.5	0.8	SE	70.3	1.4	NE
12:00 AM	67.1	1.2	SW	65.1	0.8	SW	64.2	1.3	SW	64.4	0.4	NW
14:00 AM	62.2	1.0	NE	62.1	0.7	SW	63.2	0.6	SE	63.5	0.8	NE
16:00 AM	72.0	0.6	SE	69.8	1.1	NE	70.1	1.3	SE	72.4	1.0	SE
18:00 AM	76.4	1.6	SW	74.3	1.3	SE	72.9	1.0	SW	73.0	1.4	NE

TIME	POINT 5			POINT 6			POINT 7			POINT 8		
	H	V	Dir	H	V	Dir	H	V	Dir	H	V	Dir
8:00 AM	80.1	0.8	SE	74.5	1.6	SE	74.7	0.8	WW	72.0	0.9	SE
10:00 AM	68.7	0.7	NW	69.4	1.1	NW	69.2	0.5	SW	66.4	0.5	SE
12:00 AM	63.1	2.3	NW	64.4	1.9	NW	66.2	1.6	WW	65.6	2.0	WW
14:00 AM	62.8	1.3	NW	66.6	1.8	NW	64.3	1.4	WW	62.6	2.1	WW
16:00 AM	70.0	1.2	SE	72.3	1.5	SW	72.9	1.1	SW	73.0	1.4	SE
18:00 AM	72.7	1.0	SE	72.0	0.8	SW	74.2	0.6	SW	74.8	1.1	SE

TIME	POINT 9			POINT 10			POINT 11			POINT 12		
	H	V	Dir	H	V	Dir	H	V	Dir	H	V	Dir
8:00 AM	74.1	0.8	SE	74.6	1.6	SE	72.8	1.2	SW	75.1	0.8	SE
10:00 AM	65.7	0.8	SE	65.1	2.2	SW	68.4	1.8	SW	67.2	1.1	SW
12:00 AM	65.7	1.2	NE	61.4	1.3	NW	61.0	1.6	NE	61.9	2.1	SS
14:00 AM	75.1	1.2	NE	66.6	1.2	NW	70.2	1.4	SW	72.0	1.7	SE
16:00 AM	76.1	1.1	NE	72.6	1.4	SW	69.8	1.8	NE	74.7	2.0	SE
18:00 AM	74.2	0.8	SE	73.0	1.8	SW	72.7	1.8	SW	73.1	1.4	SW

TIME	POINT 13			POINT 14			POINT 15			POINT 16		
	H	V	Dir	H	V	Dir	H	V	Dir	H	V	Dir
8:00 AM	72.7	1.3	SW	71.1	0.9	SW	75.2	1.4	SE	75.0	0.3	SE
10:00 AM	68.0	0.6	SW	66.9	1.1	SW	69.3	0.5	SW	69.2	1.4	SW
12:00 AM	60.9	1.5	SW	61.1	1.4	SW	64.3	1.2	SW	60.5	0.5	SW
14:00 AM	60.5	1.4	SW	64.6	1.9	SW	64.9	1.1	NW	62.3	1.0	NW
16:00 AM	68.0	0.8	SW	68.0	1.3	SW	70.1	0.8	SE	71.9	1.1	SE
18:00 AM	71.2	1.1	WW	71.8	1.0	SW	72.5	0.9	SW	73.1	0.5	SW

TIME	POINT 17			POINT 18								
	H	V	Dir	H	V	Dir						
8:00 AM	75.5	1.7	SE	74.6	0.8	SW						
10:00 AM	71.3	1.3	SW	66.8	0.8	SW						
12:00 AM	65.3	1.4	SW	62.9	1.1	NW						
14:00 AM	66.6	1.6	SW	62.1	1.3	NW						
16:00 AM	60.5	1.2	NW	62.6	0.7	NW						

18:00 AM	70.2	1.3	SW	73.0	0.8	SW						

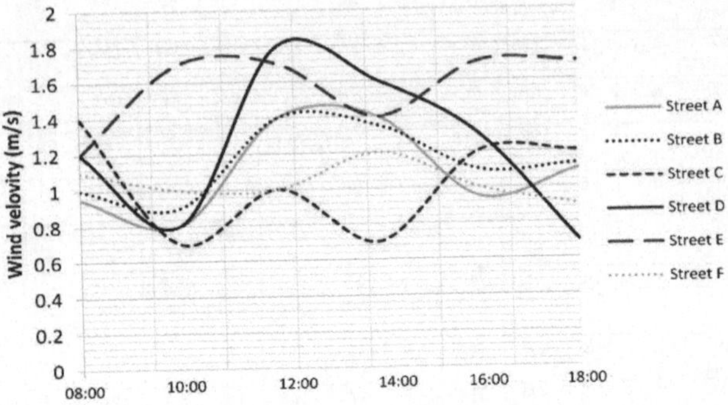

Fig. 34. Section Graphical comparison of the wind velocity across the six streets (A,B,C,D,E,F).

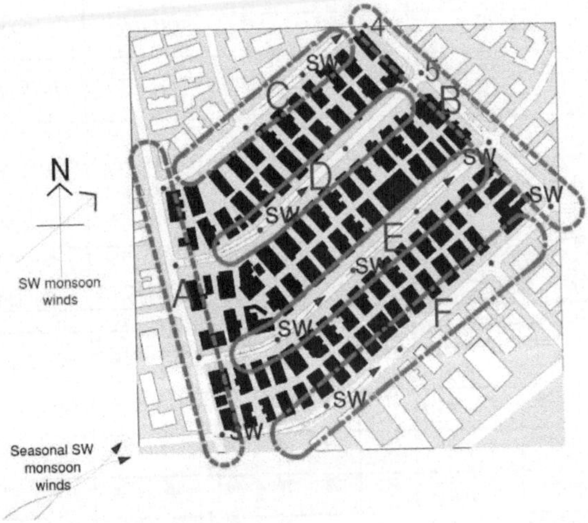

Fig. 35. The Southwesterly Monsoon seasonal wind effect in the streets during the period of study

In order to improve wind driven ventilation in such cluster settlements, understanding the airflow and other microclimatic characteristics within the streets is paramount for developing settlements that will possess energy saving considerations and human comfort. It is evident that the rapid expansion of cities creates huge areas of increased drag, with roughness and thermal forces that may significantly impact on the airflow patterns. Therefore, it is vital to provide evaluation systems with appropriate description of airflow, thermal conditions and natural ventilation outlooks within tropical regions in collaboration with meteorological frameworks that will incorporate predictions of airflow patterns on urban scales.

Also, with its characteristic climatic conditions prevalent in high temperatures and humidity, tropical cities in regions like the southeastern Nigeria, should aim towards the reduction of barriers and elements that impede on the airflow for natural ventilation. Furthermore, the significant knowledge of the wind field elements like wind velocity and direction, as well as the temperature distribution within streets levels, will enable building and settlement designers make realistic assumptions in determining the street and block configurations, building orientation decisions and appropriate sizes of openings and fenestrations on building facades, which will boost the extent of airflow for cooling and air quality purposes in dense cluster housing schemes.

CHAPTER NINE

Design Strategies for Integrating Airflow from the Urban Environment into Indoor Spaces for Effective Natural Ventilation

The process of building designs requires a clear understanding of the behaviors of airflow in and around building spaces. The understanding of the wind flow patterns and natural ventilation principles in both indoor and outdoor environments are essential for the air quality performance. Apart from the simple use of building apertures (doors and window openings etc) for introducing natural wind driven airflow for building ventilation, there are arrays of techniques for integrating natural air flow for the purposes of indoor air quality improvements and occupants thermal comfort. This section presents different techniques of utilizing natural ventilation in building design. It starts with the discussion of major modern/contemporary buildings that were specifically designed by architects to adopt the natural ventilation principles. It goes further to present design ideas that can enhance this process of natural ventilation using the building elements.

The significance of this discussion is to essentially provide students and designers with the adoptable ideas and concepts that are available for natural ventilation application in modern times.

Modern and Contemporary Buildings designed with Natural Ventilation Concepts

i. Startup Lions Campus Kenya (built by Kéré Architecture)

Designed to function as an information and communication technology campus within the banks of Lake Turkana Kenya, this building is seen as the most modern application of the Stack effect (Chimney effect) concept for Natural ventilation. The design demonstrated the designer's objective to reduce the energy loads within the building and making it more climate-friendly, cost effective and wholesomely sustainable. On the particular area of the natural ventilation concept, this was justified by the wind flow and temperature variations attainable within and around the building fabrics. With the intention for reducing its energy consumption, the intended passive ventilation techniques would enhance the building's indoor air quality as well as its thermal comfort conditions for the occupants. The building was intended to replicate the natural terrain features of the towering termite colony mounds within the region. Fig. 36 to 38 presents the design concept for the Startup Lions campus building.

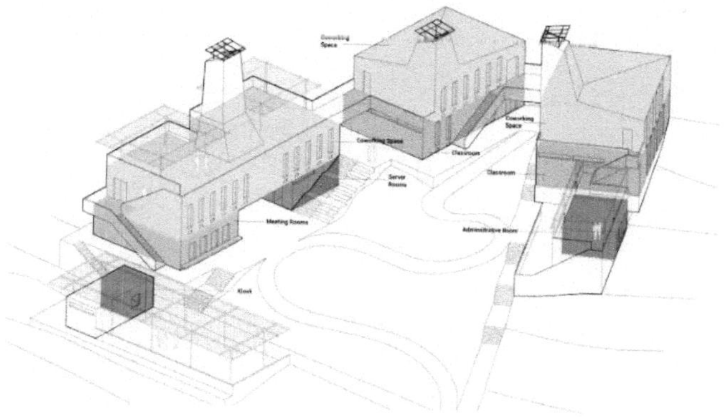

Fig. 36. Schematic layout of the Startup Lions Campus in Kenya[61].

[61] *Kéré Architecture, 2021*

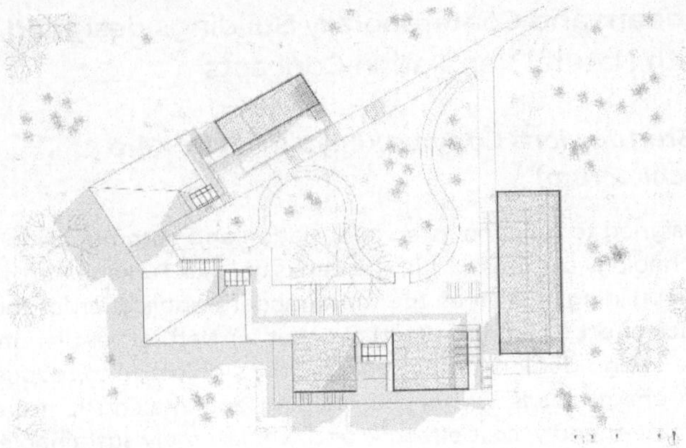

Fig. 37. Site aerial view of the Startup Lions Campus in Kenya (Source: Kéré Architecture, 2021).

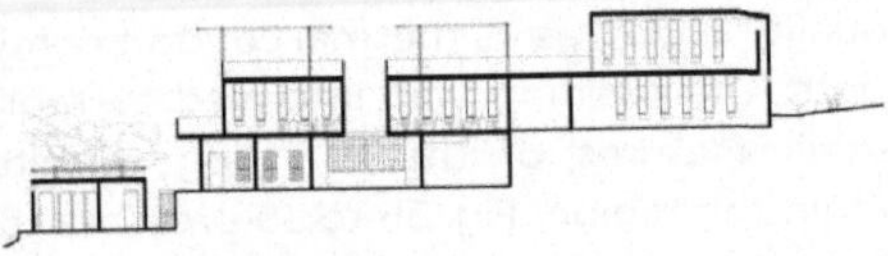

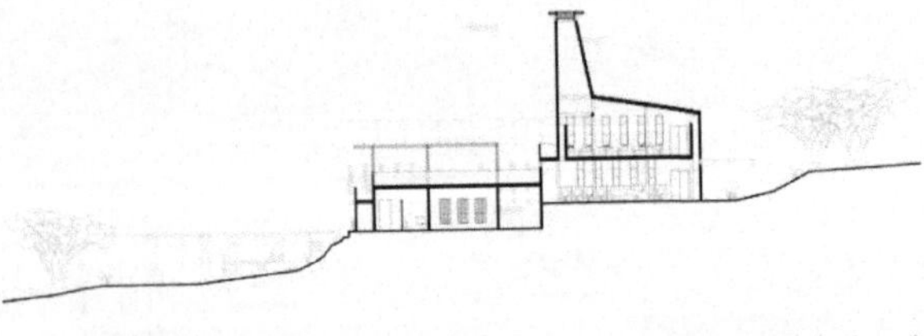

Fig. 38. Typical section of the Startup Lions Campus in Kenya (Source: Kéré Architecture, 2021).

Natural Ventilation Process Adopted in the Startup Campus Building:

The concept adopted by the designer permits the control and lowering of high temperatures within the campus as well as creates a unique landmark feature in the entire landscape. The building consists of large window openings along the building walls as well as the introduction of the "chimney type ventilation towers" in specific centers of the campus. The building's tall ventilation towers were designed to create stack effect through buoyancy conditions intended to naturally cool the main working spaces by extracting warm air upwards, while fresh air is introduced through specially designed lower level inlet openings. Open shafts were created strategically between floor to enable the unobstructed flow of hot air from the ceilings of the ground floor spaces to the chimney towers as shown in fig. 39.

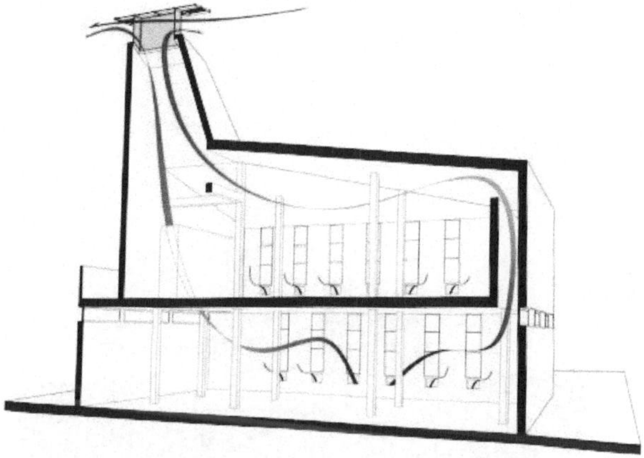

Fig. 39. Illustration of the Natural ventilation process using Stack effects in the Startup Lions Campus in Kenya (Source: Kéré Architecture, 2021).

Fig. 40. Approach view of the Startup Lions Campus in Kenya (Source: Kéré Architecture, 2021).

Fig. 41. View of the Ventilation Towers and large windows in the Startup Lions Campus (Source: Kéré Architecture, 2021).

ii. The "Box Window" Passive Ventilated House Abuja, Nigeria (built by the author)

Housing design in tropical regions of Africa necessitates that windows are created for natural ventilation and aeration of indoor spaces. However, the menace of malaria carrying mosquitos makes natural ventilation more risky than comforting. The Box Window house was created with the concept of Cross Ventilation, where natural wind driven forces move airflow across the building spaces through Box-type windows created on the building walls[62]. Most residential houses produced in Africa are made up of walls constructed of materials like concrete bricks, mud or clay bricks and adobe, with little fenestration openings across the walls owing majorly to the need to keep off external elements and for security needs as well. This is further compounded by the regular closure of doors and windows in the evening hours, resulting in uncomfortable indoor thermal and humidity conditions. These uncomfortable thermal and air-quality conditions are caused by lack of active airflow and ventilation into and out-of the houses, resulting from the action of shutting the windows and doors in order to keep the external disease causing vectors (like mosquitoes) out of the homes. According to studies, it is noted that even with the use of mosquito netting as window screens, the conventional designs of window netted screens often necessitates that these screens are first opened briefly in order to access the windows. These brief moments required to open the screens in order to gain access to the window controls is enough time for the influx of the amount of disease carrying mosquitoes and disease vectors into the homes. Likewise, owing to the fact that the challenge of healthy housing in sub-Saharan Africa doubles between the menace of malaria carrying mosquitoes and the attainment of quality indoor airflow for natural ventilation and cooling purposes, the importance of constant window opening

[62] *Box Window house is an original prototype family home designed and built by the author Akubue Jideofor Anselm in Abuja, Nigeria.*

cannot be overemphasized[63]. Therefore, this brief interval witnessed in the operation of conventional windows is identified as a flaw in the use of netted mosquito screening as solution for mosquito control and improvement of natural airflow into building spaces in this region. Images of the conventional window operating systems in Nigeria are presented in fig. 42 to 45.

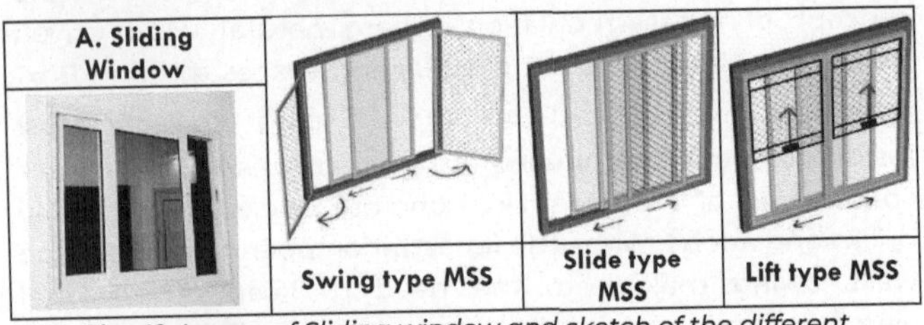

Fig. 42. Image of Sliding window and sketch of the different adoptable options of Mosquito screening system (MSS) with their opening modes

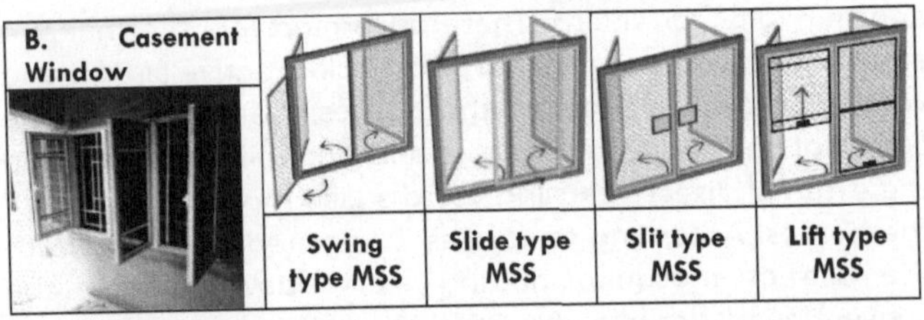

Fig. 43. Image of Casement window and sketch of the different adoptable options of Mosquito screening system (MSS) with their opening modes

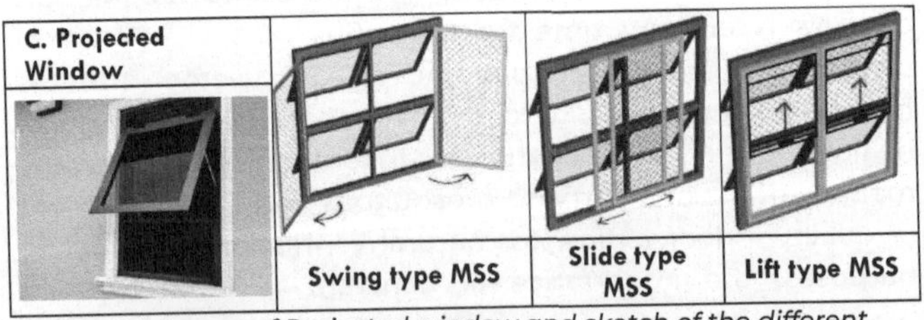

Fig. 44. Image of Projected window and sketch of the different adoptable options of Mosquito screening system (MSS) with their opening modes

---

[63] *Akubue, Jideofor. (2024). Window Design for Mosquito Control: An Architectural Solution for Reducing Malaria Burden in Tropical African Homes. Journal of Architectural Environment & Structural Engineering Research.*

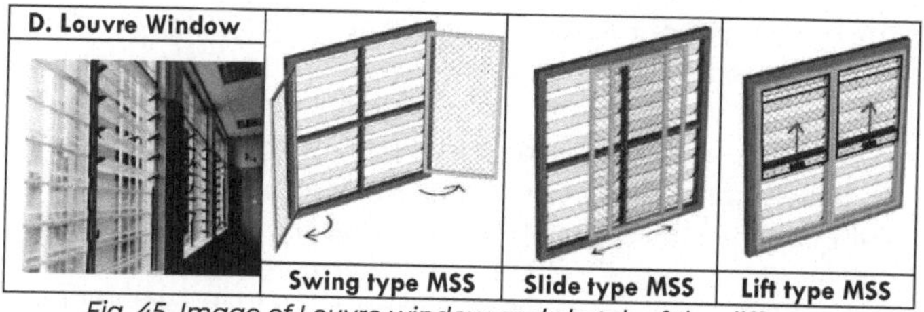

Fig. 45. Image of Louvre window and sketch of the different adoptable options of Mosquito screening system (MSS) with their opening modes

Owing to this impasse between mosquito control and attainment of full and efficient natural ventilation in this region, the development of a more effective architectural design targeted at minimizing or eliminating this interlude period is created in order to achieve both mosquito free indoor environments as well as improving the potential of constant natural ventilation. The use of the optimized Box window option is presented as an architectural solution as described in Fig. 46.

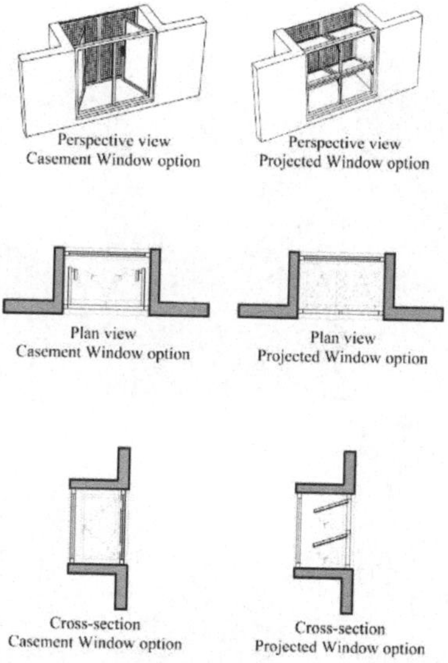

Fig. 46. Architectural design of the Optimized Mosquito screening system (OMSS), applicable in all window types including the Casement and Projected window options, (source: author's conceptual design sketches).

The design of Optimized Mosquito screening system consists of a 2 layered system comprising of the main window panel and the screen. Unlike the conventional systems which are joined with each other, the optimized design separates the panel with an air space to accommodate the 90o outwards swing of any conventional window type like the casement or projected window type. Furthermore, unlike the conventional window MSS architecture which placed the netted screen before the main window panels, the optimized (Box window) MSS architecture places the window panels before the netted screen. By this application, this design would not need to operate the netted screen in order to access the operating devices of the main window panels. This virtually eliminates the interval of exposure to mosquitos and other vectors, while giving full access to natural airflow.

For purpose of affordability of the design in typical residential homes in the region, the construction of the optimized design employed the use of lightweight composite materials comprising of sandwiched polystyrene boards in sandcrete mortar as seen in Fig. 47.

Fig. 47. Pictures of different stages of construction of the Box-window MSS using lightweight materials in the prototype house design

Natural Ventilation Process Adopted in the Box-window house:

Naturally, airflow for ventilation is created by differences in air pressure across the building. Notably, building walls impede on this airflow thereby creating wind pressure differences that occur between the windward and leeward walls of the building. Study indicates that the wind pressure on the windward surface of a building intended to drive natural ventilation across wall openings is the sum of both the static and dynamic pressures[64]. Dynamic pressure at the windward openings of buildings that are cross-ventilated is known to increase with increase in size of openings. The ratio of the opening is at a ratio of "*area of wall openings to area of walls*". In cross-ventilated spaces study also identified that the normal wind indoor pressure coefficients is about 0.6, however with inclined frequency, whereby breeze goes through the space and exits through side wall, the pressure coefficient falls to about 0.2. Measurements at building eaves levels show that the effective pressure difference is often highest with around 1.4 times that of dynamic pressure in case of rectangular buildings with all wall openings set at 15% to 20% of wall area[65]. This indicates that the average wind speed that go through wall openings possess the capability to increase up to 18% more than local wind speeds. This is a significant effect in the design for wind driven natural ventilation. In the case of the Box-window house design, the wall openings were designed to cover around 25% of the building walls. This large area of openings was intended to maximize the potential of increasing the wind speeds for natural ventilation.

Secondly, the major objective of the designer of the Box-window house was energy saving. The designer intended to maximize natural ventilation for indoor cooling by reducing dependence on mechanical ventilation by almost 80%. For this purpose the design observed all the passive ventilation consideration which foremost included the appropriate

---

[64] *Aynsley, Richard. (2014). Natural Ventilation in Passive Design. Environment Design Guide. JSTOR.*

[65] *Aynsley, R.M. (1977). Architectural Aerodynamics. Architectural science series. Applied Science Publishers,*

orientation of indoor room spaces to maximize the use of the windows distributed evenly across the exterior walls for the purposes of capturing outdoor airflow as seen in figure 48 to 50. Other passive ventilation strategies adopted includes:

- Cross ventilation of room spaces with inlet and outlet apertures
- Adoption of casement windows to enable 100% aperture access to wind flow through the windows
- Active window design to enable 24 hours access to airflow without operational obstruction.

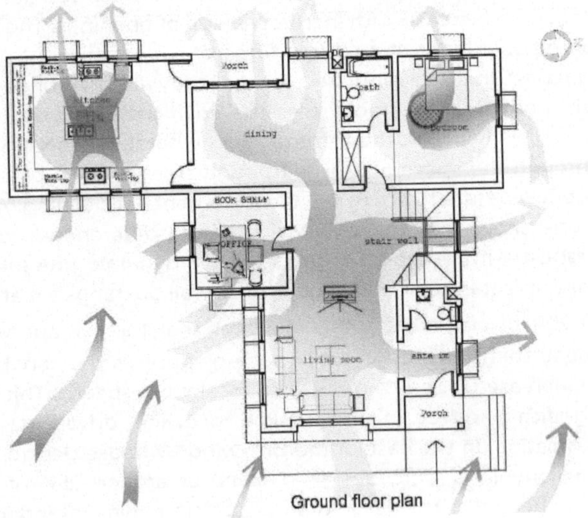

Fig. 48. Ground floor layout showing the organization of interior spaces for maximizing access to wind flow for cross ventilation of indoor spaces in the Box-window house

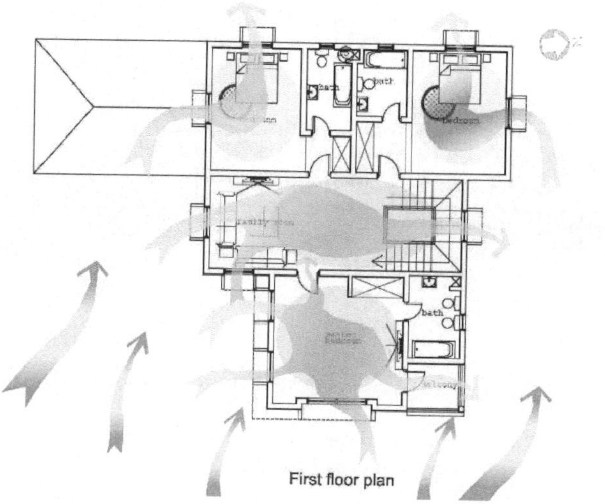

Fig. 49. First floor layout showing the organization of interior spaces for maximizing access to wind flow for cross ventilation of indoor spaces in the Box-window house

Fig. 50. X-Ray view of the Box-window house showing the effects of wind flow and cross ventilation through the windows

Fig. 51. Exterior view of the Box-window house (Source: Author's pictures of the house, 2024).

(a)

(b)

(c)

(d)

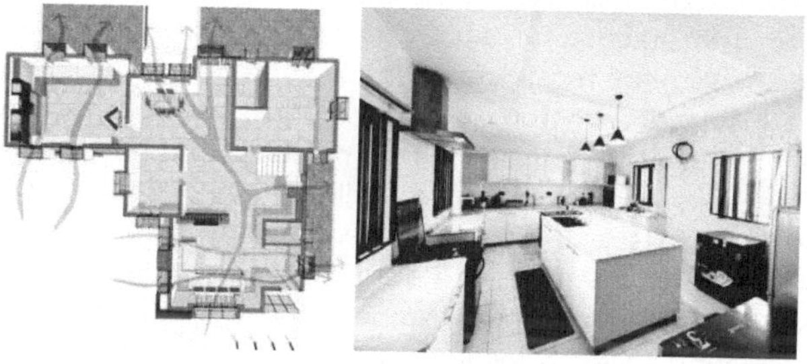

(e)

Fig. 52. Interior views of the Box-window house (a,b,c,d,e) showing the extents of cross ventilation (Source: Author's pictures of the house, 2024).

Design Strategies for Enhancing Building Cooling using Natural Ventilation Concepts

This section presents design ideas that can be adopted in building designs for the purpose of improving indoor cooling by utilizing natural ventilation processes accessed through exclusive building elements other than windows and doors. Recall that cross ventilation concepts are applicable only via openings (like doors and windows) as seen in the Box-window house, just as the stack effects or buoyancy concepts of ventilation are applicable via atriums and shafts (with inlet and outlet/exhaust systems), as seen in the Startup Lions Campus. However, this section presents the application of passive ventilation ideas for insulating building facades (walls).

It is noteworthy that the function of building facades includes the regulation of heat generated through solar radiation. It also includes moderating the ranges of thermal transfer from the exterior to the interiors of building. This primarily functions of building facades promote energy conservation, hence making the art of façade design a vital energy efficient option in sustainable building. The use of building facades as barrier-elements for thermal insulation has been in practice for ages. Concepts like the multi-skin façade system are most prominent as strategies for building space insulation, whereby double or more layers of façade covers are used to reduce the thermal loads and heat gains from the exterior environments. While the use of double skin (glazed) facades are more effective in colder climates, the introduction of a Ventilated Cavity wall system which uses passive ventilation concepts can be of same significant value for insulating building facades and improving indoor space cooling in tropical climates.

The Ventilated Cavity Wall Concept

The concept of a ventilated cavity wall is presented here as an option that functions as a barrier used to trap the excessive heat that is naturally radiated through the building envelope, while utilizing ventilation of the cavity space in its double layer to control the heated façade through aeration of the cavity space. This concept combines the Bernoulli's principles on energy conservation which utilizes the airflow effects experienced in stack ventilation, where outdoor air which is above the ground moves quicker than the air that is lower, thus maintaining a lower pressure. The resultant lower pressure is needed to suck fresh air through the cavity system thus ventilating the cavity system and retaining a lower temperature within it. The potentials of stack-effect ventilation in this case can be estimated during the design process as follows[66]:

$V = 60\ K\ A\ [g\ h\ (Ti - To)/Ti]0.5$ *Eq.6*

Where;
V is the estimated air flow rate (cfm);
K is a factor that accounts for orifice characteristics (assumed = 0.65);
A is the smallest value of inlet, stack "throat," or outlet areas (sq m);
g is the gravitational constant (9.81m/sec^2);
h is the stack height (m);
Ti (inside) is the higher of two differential air temperatures (°C);
To (outside) is the lower of two differential air temperatures (°C).

The major conditions that affect the rate of stack-effect Ventilation includes:

- Area of openings (inlet, outlet, and stack size), with the smallest of these three areas ruling.
- Difference in elevation height between the inlet and the outlet.
- Difference in dry-bulb air temperature between the stack inlet and outlet.

Out of these three conditions mentioned above, the height and area components are the most responsive to building design decision making, others are fundamentally influenced by the microclimatic conditions within the building's environment.

[66] *Wahab, I.A., Ismail, L.H. (2012). Natural Ventilation Approach in Designing Urban Tropical Houses. Conference Proceeding of International Conference of Civil and Environmental Engineering for Sustainability.*

In the application of the stack-effect principles, the ventilation rate is calculated as directly proportional to the area (m^2) of the inlet as well as the square root (√) of the height difference between the inlet and outlet openings, as seen in the equation below[67].

$V = KA_i\,(hd_t)\,0.5\ m^3s^{-1}$ *Eq.7*

Where;
V = rate of ventilation in m^3s^{-1}
K = Correction factor, which varies with the ratio of outlet to inlet areas.
A_i = area of inlet.
h = height difference between inlet and outlet.
d_t = temperature difference.

For the purpose of this study, a physical model was developed. It comprises of two exterior wall panels with the cavity (air corridor) for ventilation of the cavity air. The study model is 4.2m high and $12m^2$ in area. The building box was made with two layers of conventional bricks for the inner wall which is 150mm thick and the outer wall made of sandwich polyurethane-foam insulation wall panel of about 75mm in width. The sandwich panel comprises of a 50mm thick polyurethane-foam board, which is coated on both sides with a 12.5mm thick cement mortar. The cavity space between the double layers is 700mm and is finished with a reflective coating to enhance the thermal insulation of the cavity space. A glazed window is added on the exterior skin for lighting and ventilation when required and also fitted with a glazed door in the interior skin that opens into the central cavity unit. These cavity units can be opened at the floor level and roof level for the purposes of passive ventilation of the cavity air through buoyancy or stack effects as seen in fig. 53.

---

[67] *Markus T. A., Morris E. N. (1980). Buildings, Climate and Energy. London: Pitman Publishing Limited*

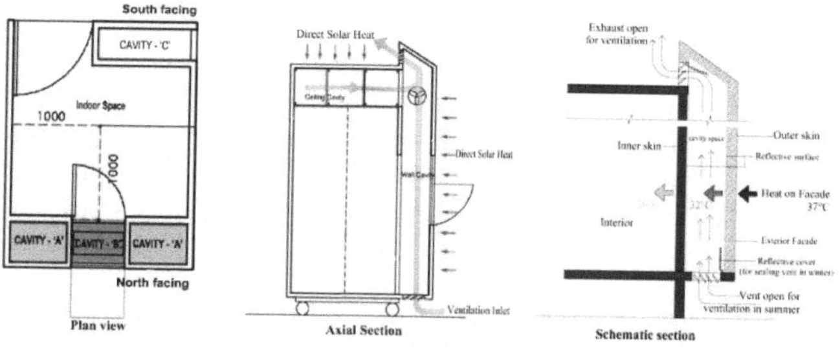

Fig. 53. Details of the Ventilated Cavity Wall design model

Fig. 54. Images of the Ventilated Cavity Wall experimental model. (Source: Author's pictures of the model, 2008).

The study of the model identified the extent of the impact the ventilation of the cavity wall has on the Ventilated cavity wall model. As the external conditions (outdoors) heats up, the outer skin of the double skin wall heats up as well. However, owing to the presence of the heat resistant insulation material, which is a polyurethane foam board with an R-value of 5.8, the thermal gain (through solar radiation) is reduced. Furthermore, the cavity air space in between the double skins which is ventilated from below (inlet region) passively ventilates the warmed-up air inside the cavity space (as illustrated in fig. 55). Through the process of air filtration (via buoyancy and stack-effects), the air flows upwards and exits via the outlet opening at the upper end of the wall. This process minimizes the overall heat gain by the exterior layer before getting to the internal layer of the skin, thereby preventing the heat gain from getting to the internal room spaces.

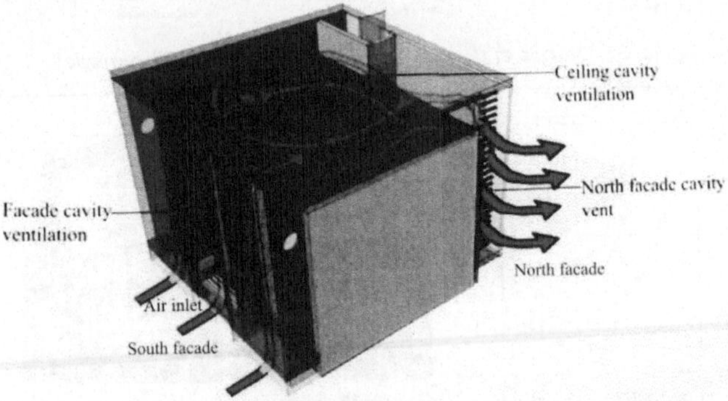

Fig. 55. An X-rayed view of the Ventilated Cavity Wall showing process of ventilation through the inlet and outlet vent points

The experiment conducted on the model is intended to identify the impact of the ventilated system when active (the cavity space is ventilated) and when inactive (closed system). As seen in fig. 56 and 57, the process of ventilation (with opening of the inlet and exhaust vents), permits airflow which reduces the cavity air temperature significantly. This could be observed at the peak of the heated conditions with the external surface indicating 39°C and the internal layer indicating 34°C. The internal cavity air temperature in this case recorded 37°C against an outdoor temperature of 40°C indicating a significant drop of about 3°C between the outdoor air temperature and the cavity air temperature. However in the case of the closed ventilation system, the external surface indicated 38.5°C and the internal layer indicating 37°C. The internal cavity air temperature in this closed vent scenario recorded 38°C against an outdoor temperature of 39°C, signifying a drop of only 1°C between the outdoor air temperature and the cavity air temperature.

The differences observed in the open vent (ventilated wall) with a 3°C improvement and the closed vent wall with only a 1°C improvement goes to show the significant impact of the ventilated cavity wall when active.

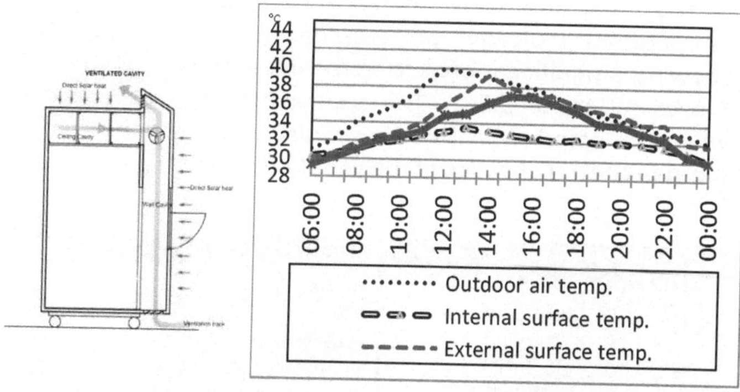

Fig. 56. Analysis of the ventilated cavity wall model (north facing) when ventilated with open inlet and exhaust

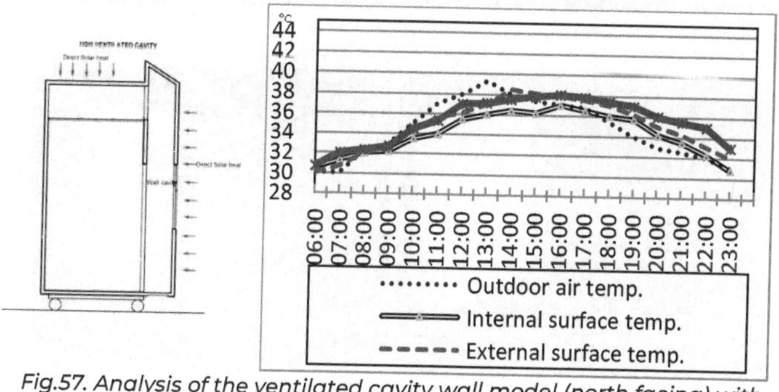

Fig.57. Analysis of the ventilated cavity wall model (north facing) with Closed-vent and exhaust

Using the key data optioned from the measurements, the air temperatures of the inlets into the cavity space, inlet air velocities and the external wall surface temperatures for the north and south facing walls were inputted into the

PHOENIX-VR CFD modeling tool for the simulation of the thermal conditions. Despite an external surface temperature of about 38°C and 40°C, the simulation results revealed a significant reduction in internal surface temperature of the ventilated cavity walls to an average of 33°C. Similarly, the ventilated cavity space air temperature recorded a value of 36°C, which signifies the optimistic impact of the ventilated cavity space in reducing the heat gained via the building's facades. On the other hand, final indoor ambient air temperature indicated an average value of about 31°C, showing a significant drop from the initial air temperature of the cavity space (36°C). The results from the experiment when compared to that of the simulation, demonstrates the thermal control qualities of the ventilated cavity wall system.

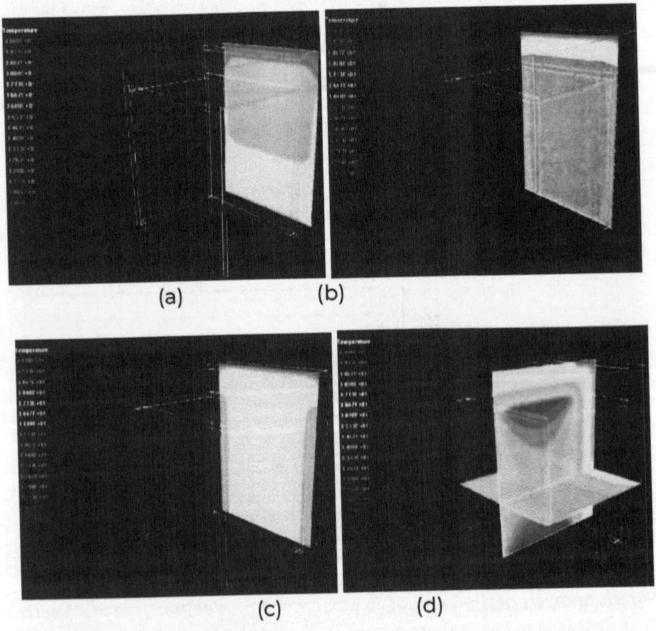

Fig. 58. Results from the simulation of the thermal conditions in the north facing ventilated cavity wall. (a) Surface temperature of the external wall, (b) Surface temperature of the internal wall, (c) Air temperature of the Cavity space, (d) Indoor room air temperature.

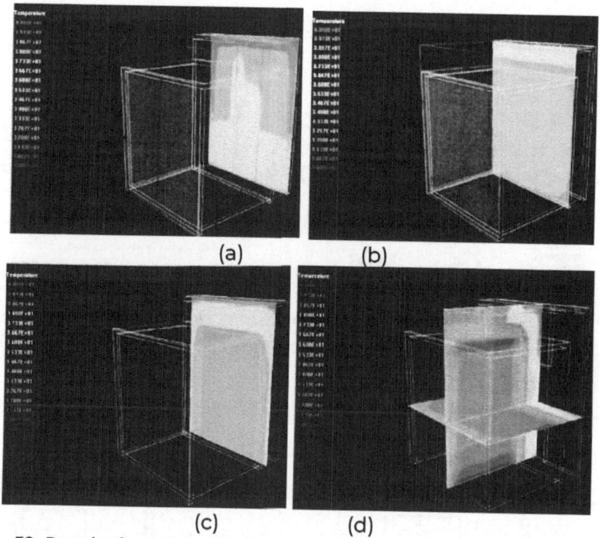

Fig. 59. Results from the simulation of the thermal conditions in the south facing ventilated cavity wall. (a) Surface temperature of the external wall, (b) Surface temperature of the internal wall, (c) Air temperature of the Cavity space, (d) Indoor room air temperature.

Following the results of this study, various configurations can be adopted for application of the ventilated cavity wall in building design. The foremost consideration in the decision for its application is the orientation of the building, in order to maximize the use of natural wind driven ventilation for activating the cavity air flow. Other options include the addition of Hybrid vents at the exhaust points of the façade as to increase the air flow rate in the ventilation process. Fig. 60 presents the configurations that can be adopted in the design of buildings using this façade type.

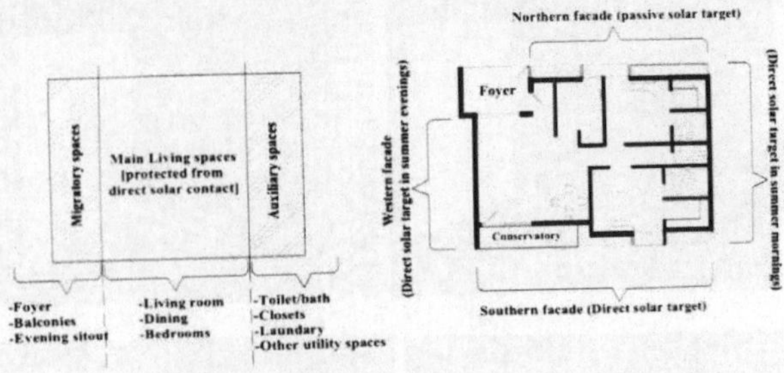

Fig. 60. Schematic concept for spatial distribution within an apartment unit adopting the Ventilated cavity wall façade system

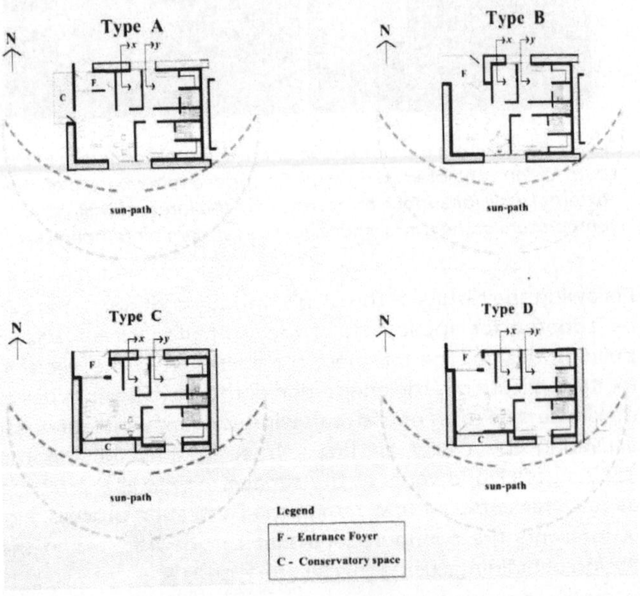

Fig. 61. Different configurations for the application of the ventilated cavity wall in an apartment design

Epilogue

Natural ventilation is summarized in this book as the use of natural forces like winds and thermal buoyancy (owing to differences in indoor and outdoor air density) to draw air through apertures created within the building envelopes. These apertures as analyzed in this book include the likes of windows, doors, vents, chimneys, wind towers and cavities. These natural drivers of ventilation are discussed further as a way of broadening students' understanding of the principles of wind-flow for the purposes of effective application of natural ventilation in building designs.

This book comprise of nine chapters namely:

1. Principles of Ventilation
2. Types of Ventilation
3. Natural Ventilation Principles
4. Application of Natural Ventilation in Building Design
5. Principles of Airflow for Natural Ventilation in the Built Environment
6. Airflow Regimes
7. Impacts of Airflow Regimes on Buildings
8. Analysis of Wind flow Patterns for activating Natural Ventilation in the Urban
9. Design Strategies for Integrating Airflow Patterns from the urban environment into Indoor Spaces for Effective Natural Ventilation

While chapter one narrates the general principles of ventilation, chapter two discusses the basic types of ventilation. Chapters three and four discuss the natural ventilation principles and their application in building design. In chapter five, this book outlines the principles of airflow for natural ventilation in the built environment, while describing the layers of airflow regimes in chapter six. Chapter seven presents the behaviors and impacts of airflow on different layers of the building canopy, while chapter eight presents some analysis of wind-flow patterns on building geometry and street configurations as means of evaluating the behaviors of wind-flow patterns in activating natural ventilation within Urban environments. The discussions ended in chapter nine with the presentation of design strategies for integrating these wind-flow patterns within building spaces for effective natural ventilation process.

The significance of the studies conducted in this book lies with the tinkering of student's knowledge of the wind field elements and behaviors. This knowledge may enable them make informed decisions during building and settlement designs while improving their quest for realistic assumptions in determining the impacts of natural ventilation principles on street and building configurations, building orientation decisions as well as appropriate sizes of openings and fenestrations on building facades; all geared towards boosting the extent of airflow for cooling and air quality purposes.

In conclusion, it is evident that the rapid expansion of cities creates huge areas of increased drag, with roughness and thermal forces that may significantly impact on airflow patterns. Therefore, it is vital to conduct such studies as discussed in this book in order to generate localized evaluation systems with appropriate description of airflow, thermal conditions and natural ventilation outlooks within tropical regions in collaboration with meteorological frameworks that will integrate predictions of airflow patterns on urban scales.

Bibliography

1. Abd W. I., Ismail, L. & Kadir, A. (2016). Opening Design and Position Effect on Building Natural Stack Effect and Cross Ventilation. International Journal of Engineering and Technical Research. 5. 13-22.
2. Adam, J. P. (2005): Roman building: materials and techniques. Routledge.
3. AIVC (1996). "GV: A Guide to Energy Efficient Ventilation
4. Aiyer, P. (2021). Wind Patterns in Built Environment. International Journal Of Research And Analytical Reviews. 11. 855-862. 10.1729/Journal.41641.
5. Akubue, J. A. (2019). Effects of Street Geometry on Airflow Regimes for Natural Ventilation in Three Different Street Configurations in Enugu City. Strategies of Housing Design, Intechopen, 2, 11-26.
6. Akubue, J.A. (2024). Window Design for Mosquito Control: An Architectural Solution for Reducing Malaria Burden in Tropical African Homes. Journal of Architectural Environment & Structural Engineering Research. 7. 10-19.
7. Al-Sallal A., & Al-Rais L. (2012). Outdoor airflow analysis and potential for passive cooling in the modern urban context of Dubai. Renewable Energy, 38, 40-49.
8. Ali-Toudert, F., & Mayer, H. (2006). Numerical study on the effects of aspect ratio and orientation of an urban street canyon on outdoor thermal comfort in hot and dry climate. Buildings and Environment; 41:94-108
9. Andreou, E. & Axarli, K. (2012). Investigation of urban canyon microclimate in traditional and contemporary environment. Experi-mental investigation and parametric analysis. Renewable Energy, 43, 354-363
10. ASHRAE (2007). Handbook of HVAC applications. Atlanta, USA: American Society of Heating Refrigerating and Air-Conditioning Engineers Inc. Health care facilities.
11. Aynsley, R.M. (2014). Natural Ventilation in Passive Design. Environment Design Guide. JSTOR, p. 1–16.
12. Aynsley, R.M. (1977). Architectural Aerodynamics. Architectural science series. Applied Science Publishers, Melbourne, W.H.
13. Blocken, B., Stathopoulos, T., Carmeliet, J. & Hensen, J. (2011). Application of computational fluid dynamics in building

performance simulation for the outdoor environment: An overview. Journal of Building Performance and Simulation, 4, 157–184

14. Britter, R.E., & Hanna, S.R. (2003). Flow and Dispersion in Urban Areas. Annual Review of Fluid Mechanics, 35, 469–496.
15. Bourbia, F., & Boucheriba, F. (2010). Impact of street design on urban microclimate for semi-arid climate (Constantine). Renewable Energy, 35, 343-347
16. Boutet, T.S. (1987). Air Movement; A Manual for Architects Controlling and Builders. Mc-Graw Hill: New York.
17. Bozonnet, E. (2006). Modelling air flows around buildings in urban environment. International Workshop on Energy Performance and Environmental 1.
18. Buildings Department (2014). Guidelines on Design and Construction Requirements for Energy Efficiency of Residential Buildings. Hong Kong Government, Buildings Department, Code and Design Manuals.
19. Cambridge Advanced Learner's Dictionary, third edition (2008). Cambridge university press.
20. Caracuta, V., & Fiorentino, G. (2013). Wood for fuel in Roman hypocaust baths: new data from the late-Roman villa of Faragola. (SE Italy). Saguntum 11. Sagvntum.
21. Clancy, L. J. (1975). Aerodynamics. Wiley.
22. Cook, N.J. (1985) The Designer's Guide to Wind Loading of Building Structures. Part 1: Background, Damage Survey, Wind Data and Structural Classification. Building Research Establishment. Garston and Butterworths, London
23. Danielis Bernoulli (1738). Hydrodynamica, sive de viribus et motibus fluidorum commentarii : opus academicum ab auctore, dum Petropoli ageret, congestum. https://doi.org/10.3931/e-rara-3911
24. Etheridge, D.W., & Sandberg, M. A simple parametric study of ventilation, Building and Environment, Volume 19, Issue 3, 1984
25. Etheridge D.W, & Sandberg, M. Building ventilation; theory and measurement. Chichester, UK, John Wiley & Sons, 1996.
26. Fathy, H. (1986) Natural energy and vernacular architecture, principles and examples with reference to hot arid climates. The University of Chicago Press, Chicago
27. Fernando, H. J. S. (2010). Fluid mechanics of urban atmospheres in complex terrain. Annual Review. Fluid Mechanics, 42, 365–389.
28. Feustel, H.E. (1999). COMIS—an international multi zone air-flow and contaminant transport model, Energy and Buildings. 30(1), 3-18
29. Forbes (1966). Studies in Ancient Technology. BRILL. p. 54–55.

30. Fuad, H.M. (1996). Thermal comfort and building design in the tropical climates. Energy and Buildings, 23, 161-167

31. Garfinkel, Y. (1993). "The Yarmukian Culture in Israel". Paléorient, 19.1:115 – 134.

32. Geros, V., Santamouris, M., Tsangrassoulis, A. & Guarracino, G. (1999). Experimental evaluation of night ventilation phenomena, Energy and Buildings, 29, 141–154.

33. Georgakis, C. & Santamouris, M. (2006). Experimental investigation of air flow and temperature distribution in deep urban canyons for natural ventilation purposes. Energy and Buildings, 38, 367-376

34. Ghiaus, C., & Allard, F. (2005). "Natural Ventilation in the Urban Environment: Assessment and Design", Buildings Energy and Solar Technology. London: Earthscan

35. Givoni, B., Noguchi, M., Saaroni, H., Pochter, O., Yaacov, Y. & Feller, N. (2003). Outdoor comfort research issues. Energy and Building, 35, 77-86

36. Gomez, F., Perez, A., Valcuende, M., & Matzarakis, A. (2013). Research on ecological design to enhance comfort in open spaces of a city (Valencia, Spain). Utility of the physiological equivalent temperature (PET), 57, 27-39

37. Grimmond, C.S.B., & Oke, T.R. (2002). Turbulent heat fluxes in urban areas: Observations and a local-scale urban meteorological parameterization scheme (LUMPS). Journal of Applied Meteorology. Climatology. 41, 792–810.

38. Grimmond, C.S.B., Roth, M., Oke, T.R., Au, Y.C., Best, M., Betts, R., Carmichael, G., Cleugh, H., Dabberdt, W., Emmanuel, R., Freitas, E., Fortuniak, K., Hanna, S., Klein, P., Kalkstein, L.S., Liu, C.H., Nickson, A., Pearlmutter, D., Sallor, D., Voogt, J. (2010). Climate and More Sustainable Cities: Climate Information for Improved Planning and Management of Cities (Producers/Capabilities Perspective). Procedia Environmental Sciences; 1:247-274

39. Huang, L.M., Li, H.T., Zhu, D.H. (2008). A fieldwork study on the diurnal changes of urban microclimate in four types of ground cover and urban heat island of Nanjing, China. Buildings and environment; 43: 7-17

40. Hunt, J. (1975). Fundamental studies of wind flow near buildings. In Models and Systems in Architecture and Building—LUBFS; Conference Proceedings No. 2; Construction Press: Berlin, Germany.

41. Institution of Heating and Ventilation Engineers (IHVE). (1970). A Guide Book. London

42. Kim, J. J. & Baik, J. J. (1999). A numerical study of thermal effects on flow and pollution dispersion in urban street canyons. Journal of Applied Meteorology, 38, 1249–1261.

43. Kim, Y.C., Yoshida, A., & Tamura, Y. (2012). "Characteristics of surface wind pressures on low-rise building located among large group of surrounding buildings", Engineering Structures, 35,18-28

44. Mahmood, F. & Al-Ansari, T. (2023). Renewable Energy Driven Sustainable Greenhouse: An Overview. 10.1016/B978-0-323-90386-8.00073-5

45. Markus, T.A. & Morris, E.N. (1980) Buildings, Climate and Energy. Pitman Publishing Limited, London.

46. Markus, B. (2016). Review of Courtyard House in Nigeria: definitions, history, evolution, typology, and functions. AFRREV STECH: An International Journal of Science and Technology. 5, 103-117

47. Memon, R. A. & Leung, D. Y. C. (2010). Impacts of environmental factors on urban heating. Journal of Environmental Sciences China, 22 (12), 1903-1909

48. Mitchell, P. (2008). Central Heating, Installation, Maintenance and Repair. Writers Print Shop

49. Mistriotis, A., Arcidiacono, C., Picuno, P., Bot, G. & Scarascia-Mugnozza, P.A. (1997). "Computational analysis of ventilation in greenhouses at zero-and lowwind- speeds", Agricultural and Forest Meteorology. 88(1-4), 121-35

50. Nakamura, Y. & Oke, T.R. (1988). Wind, temperature and stability conditions in an east-west oriented urban canyon. Atmospheric Environment; 22: 2691-2700

51. Ng, E., Yuan, C., Chen, L., Ren, C. & Fung, J.C. (2011). Improving the wind environment in high-density cities by understanding urban morphology and surface roughness: A study in Hong Kong. Landscaping and Urban Planning, 101, 59–74.

52. Nielsen, P. V. (1995). Lecture Notes on Mixing Ventilation. Dept. of Building Technology and Structural Engineering, Aalborg University. Gul series Vol. U9513 No. 29

53. Nithya, D. S., Quaranta, G., Muscarello, V., & Liang, M. (2024). Review of Wind Flow Modelling in Urban Environments to Support the Development of Urban Air Mobility. Drones, 8(4), 147

54. Ofomata, G E. (2002). A Survey of Igbo Nations. Africana first publishers, Onitsha Nigeria, pp. xxiv-703.

55. Ojo, O. (1967). The climates of West Africa. London: Heinemann.

56. Oke, T.R. (1976). Inadvertent modification of the city atmosphere and the prospects for planned urban climates. Procedure Sympozium Meteorology Related to Urban and Regional Land-Use Planning, Asheville, N.C., World Meteorology Organization, Geneva, 151–75.

57. Oke, T. R. (1987). Boundary Layer Climates. Routledge

58. Oke, T. R. (1988). Street design and urban canopy layer climate. Energy and Buildings, 11, 103–113.

59. Oyem, A.A & Igbafe, A.I. (2010). Analysis of the atmospheric aerosol loading over Nigeria. Environmental Research Journal, 4(1), 145-156

60. Peterka, J.A., Meroney, R.N. & Kothari, K.M. (1985). Wind flow patterns around buildings. Journal of Wind Engineering and Industrial Aerodynamics, 21, 21–38

61. Pourteimouri, P., Campmans, G.H.P., Wijnberg, K.M. & Hulscher, S. A. (2022). Numerical Study on the Impact of Building Dimensions on Airflow Patterns and Bed Morphology around Buildings at the Beach. Journal of Marine Science and Engineering, 10, 13.

62. Reifsnyder, W., & Darnhofer, T. (1989). Meteorology and agroforestry. World Agroforestry Centre. Nairobi, pp. 544

63. Sakakibara, Y. (1996). A numerical study of the effect of urban geometry upon the surface energy budget. Atmospheric Environment, 30, 487–496.

64. Santamouris, M. (2001). Energy and climate in the urban environment. London: James and James Science Publishers London.

65. Saxon, R. (1994).The Atrium Comes of Age. Longman United Kingdom.

66. Sehulster, L. & Chinn, R. (2003). Guidelines for Environmental Infection Control in Health Care Facilities Recommendations of CDC and the Healthcare Infection Control Practices Advisory Committee (HICPAC). MMWR. Recommendations and reports: Morbidity and mortality weekly report. Recommendations and reports / Centers for Disease Control. 52. 1-42.

67. Srivanit, M., & Kazunori, H. (2011). The Influence of Urban Morphology Indicators on Summer Diurnal Range of Urban Cllmate in Bangkok Metropolitan area, Thailand. International Journal of Civil & Environmental Engineering, 11(5):34-46

68. Su, Y., Riffat, S., Lin, Y., & Khan, N., "Experimental and CFD study of ventilation flow rate of a Monodraught TM windcatcher", Energy and Buildings, 40, 1110-1116.

69. Todhunter, P.E. (1990). Microclimatic Variations Attributable to Urban Canyon Asymmetry and Orientation. Physics and Geography.; 11:131-141

70. Tomlinson, C. (1850). A rudimentary treatise on warming and ventilation: being a concise exposition of the general principles of the art of warming and ventilating domestic and public buildings, mines, lighthouses, ships, etc. J. Weale. Hypocaust.

71. Townsend, A.A. (1976). The structure of turbulent shear flow; Boundary layers and wall jets. 2nd edition. London: Cambridge University Press;. p. 259-333

72. Tseng, Y. H. & Meneveau, C. (2006). Parlange, M. B. Modeling flow around bluff bodies and urban dispersion using large eddy simulation. Environmental Science Technology, 40, 2653–2662.

73. Wahab, I.A., Ismail, & L.H. (2012). Natural Ventilation Approach in Designing Urban Tropical Houses. Conference Proceeding of International Conference of Civil and Environmental Engineering for Sustainability.

74. Xie, Z.T., Coceal, O. & Castro, I.P. (2008). Large-Eddy Simulation of flows over Random Urban-like Obstacles. Boundary Layer Meteorology, 129, 1–23.

75. Yang, T. & Clements-Croome, D. (2012). Natural Ventilation in Built Environment. 10.1007/978-1-4419-0851-3_488.

76. Yoshida, A., Tominaga, K. & Watani, S. (1991). Field measurements on energy balance of an urban canyon in the summer season. Energy and Buildings, 15&16, 417–23.

77. Zhai, Z. (2006). "Application of computational fluid dynamics in building design: aspects and trends", Indoor and Built Environment. 15, 305-313.